MAKING AND UNMAKING OF SAN DIEGO BAY

MAKING AND UNMAKING OF SAN DIEGO BAY

Matthew R. Kaser and Gary C. Howard

CRC Press
Taylor & Francis Group
Boca Raton London New York

CRC Press is an imprint of the
Taylor & Francis Group, an **informa** business

First edition published 2022
by CRC Press
6000 Broken Sound Parkway NW, Suite 300, Boca Raton, FL 33487-2742

and by CRC Press
2 Park Square, Milton Park, Abingdon, Oxon, OX14 4RN

CRC Press is an imprint of Taylor & Francis Group, LLC

Library of Congress Cataloging-in-Publication Data
A catalog record for this title has been requested

ISBN: 978-1-138-59676-4 (hbk)
ISBN: 978-1-032-10244-3 (pbk)
ISBN: 978-0-429-48746-0 (ebk)

DOI: 10.1201/9780429487460

Typeset in Times
by MPS Limited, Dehradun

Contents

1 California Then and Now

San Diego Bay is a beautiful natural harbor with a mild Mediterranean climate and access to great beaches, scenic mountains, and fascinating deserts. These features combine to make San Diego a highly desirable place to live and work. The Bay itself is about 20 km long and 0.5–1.5 km wide. It is quite shallow over most of its area. Its average depth is 6.5 meters, and at its deepest, it is about 20 meters. The Bay is nearly surrounded by land. It is formed by a long narrow peninsula that separates the Bay from the Ocean. At the northern end, the Bay connects to the Pacific through an opening that is less than half a mile wide. The peninsula that forms San Diego Bay protects the harbor for shipping. This world-class port is one of the busiest on the West Coast and a significant node for the cruise industry. In addition, it is a large home port and training center for the U.S. Navy, and Camp Pendleton just to the north is another large post for the U.S. Marine Corps.

The region includes more than just the Bay (Figure 1.1). It stretches from Del Mar in the north to the border with Mexico in the south and east to the Peninsular Ranges and the San Andreas fault. The Peninsular Ranges cover over 300 km from Southern California to the southern tip of Baja California. Their elevations vary from 150 meters to just over 3 km. Several rivers flow into or near San Diego Bay. Mission Bay, just north of San Diego Bay, is about 2000 acres. Juan Rodriquez Cabrillo named it False Bay in 1542. It contained lagoons, estuaries, tidal marshes, and bays. This historically shallow bay has now been modified extensively by humans and is now a major entertainment destination. Other smaller bays and wetlands surround much of the Bay. The wetlands have been greatly reduced since the arrival of Europeans to the Bay Area, but efforts are being made to replace them now.

In addition, the Bay region is highly developed and includes several cities (Figure 1.2). In 2021, the population of the metropolitan area is 3,272,000, making it the 15th largest metro area in the United States. The city of San Diego sits on the northeast side of the Bay. National City and Chula Vista are south of San Diego, and Imperial Beach is at the southern end of the Bay. The land north of San Diego is Point Loma, which hooks around the end of the peninsula that forms the Bay. Coronado sits at the end of the peninsula and houses a massive base for the U.S. Navy. The area is just north of the U.S./Mexico border and serves as a major trade and transit point between the two nations.

The natural history of the region is obviously linked to that of California (Figure 1.3. This figure will be useful throughout the book). What we know as California today was a relatively late addition to the west coast of North America.

DOI: 10.1201/9780429487460-1

FIGURE 1.1 San Diego Bay. The Bay is a beautiful natural harbor. It is about 20 km long and 0.5–1.5 km wide. It has a mild Mediterranean climate and access to beaches, mountains, and deserts (Photograph courtesy of NASA).

About 150 million years ago or so, the West Coast was roughly at Arizona, and today's California coast did not exist. The origin of this part of California lies in the middle of the Pacific Ocean and involves plate tectonics. The surface of the Earth is divided into plates that "float" on the hotter layer under the crust, the mantle. New crust is created at mid-ocean ridges, and it pushes the plates around the planet. This process was deeply involved in the development of modern-day Southern California. It began with the collision of two of the Earth's massive tectonic plates. The heavier oceanic Farallon plate was subducted by the lighter continental North American plate. As the Farallon plate dived under the other, considerable amounts of material from the floor of the Pacific Ocean were scrapped up and added to the growing coast of California. As the last portions of the Farallon plate were subducted, a few pieces broke off to open windows to part

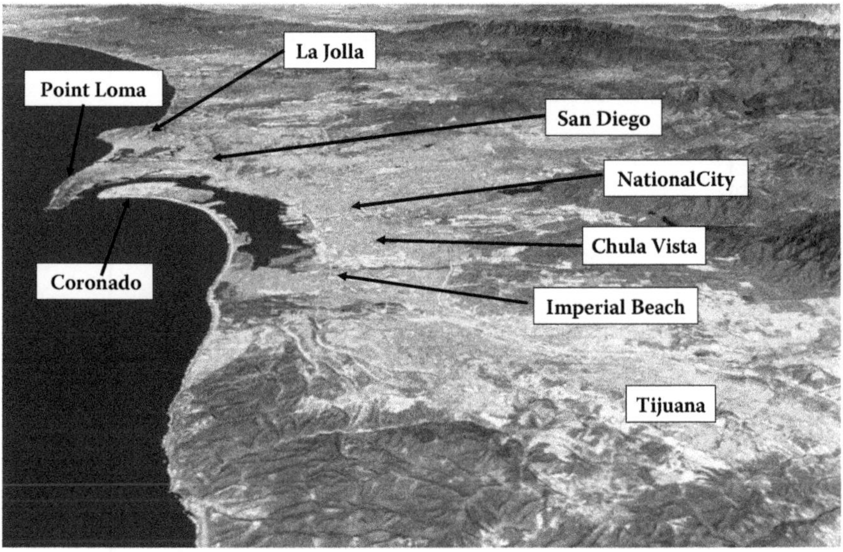

FIGURE 1.2 The Bay region is highly developed. Several major cities surround the Bay, including San Diego, Chula Vista, National City, and Imperial Beach, and Tijuana is just to the south (Photograph courtesy of NASA; cities added by authors).

of the mantle that allowed very hot rock to reach the surface as small volcanoes. Finally, with most of the Farallon plate out of the way, the Pacific plate collided with the North American plate to form the San Andreas fault. For the most part, these actions were underway when the dinosaurs roamed the Earth. Thus, the San Diego Bay region has mostly marine and only a few terrestrial dinosaur fossils.

The rocks that were scrapped up off the bottom of the ocean make up much of San Diego Bay region and of Coastal California in general. As they were added to the North American plate, they broke off into distinct sections, called terranes. Terranes are masses of rock that were accreted from the Pacific Ocean floor to the growing coast of California. The terranes consist of heterogeneous rock. Their western boundary is the San Andreas fault. The most common rocks in this Franciscan Formation are a hard sandstone called greywacke, shales, and conglomerates that have all been partially metamorphosized. Others include chert, basalt, limestone, serpentinite, and blueschists.

Part of the Southern California batholith on the western side of the San Andreas fault was detached from the mainland. In other words, it is part of the Pacific plate that is moving northward as it slips along against the North American plate. About 4.5 million years ago, that landmass had migrated northward to about the position of present-day Monterey, California.

While water is not quite so powerful a force as the tectonic plate movement, it has nevertheless had a significant role in sculpting the land of the San Diego Bay region. Rising and lowering sea levels have also influenced the size of the Bay

Cenozoic	Quarternary	Holocene	*0.012
		Pleistocene	2.58
	Neogene	Pilocene	5.33
		Miocene	23.0
	Paleogene	Oligocene	33.9
		Eocene	56.0
		Paleocene	66.0
Mesozoic	Cretaceous	Upper	100.5
		Lower	145.0

FIGURE 1.3 Geologic and biological timeline. The asterisk indicates million years ago.

itself. In fact, the Bay had alternated between wet and dry over the last 2 million years as Ice Ages came and went and sea levels rose and fell. Its evolution is a fascinating story. It involves the movements of tectonic plates, the rise and fall of oceans, the coming and going of water, and the actions of humans. It was not always here, and it will not be here in the future. Changes in the global climate resulting from the last ice age yielded lower and then higher ocean levels. In addition, water has more effects. Atmospheric rivers periodically rain on the region. They have often ended extended periods of drought. That cycle of too little water and then too much has cycled over Southern California for millennia.

The Bay has a diverse ecology and has been home to a wide variety of plants and animals since its formation. Over the millions of years, those plants and animals have evolved. Early on, the animals included many of the large mammals that characterized western North American. Those large mammals and many others have become extinct, and others have evolved to take their places. The San Diego Bay region is also part of the Pacific flyway for migratory birds. Several areas are parts of the San Diego National Wildlife Refuge Complex. The wildlife refuge protects multiple habitats, including coastal marshes and uplands, chaparral, coastal sage scrub, oak woodland, freshwater marsh, vernal pools, and breeding and nesting grounds for migratory and resident birds.

Approximately 10,000 years ago, Native Americans living in the San Diego Bay region would have noticed changes to their environment. They arrived during the last Ice Age when sea levels were lower than today, and the coast was 15–50 km west of where it is today. When that coldest part of the Ice Age ended, the sea level began to rise. Year after year, the tides were higher than those before. The ocean was creeping further and further onto the land. Those early Native Americans likely saw the water return to the San Diego Bay.

In more recent years, human activity has greatly influenced the Bay, and much of that change has not been for the better. Humans built several major cities and filled significant parts of the Bay. The region continues to increase in population, and the additional people put more stress on the housing, transportation, and other systems in the region. Bridges cross the bay and rivers and major highways connect the cities. The natural systems have too often been the losers since humans entered the Bay region in large numbers. Unfortunately, human activities have introduced quite a number of invasive species that have disrupted natural populations and, in some cases, led to their extinction. The drainage of pollution into the Pacific Ocean reaches the beaches around San Diego, and the solution to that problem is complicated by an international border.

There is some good news. In the last few decades, people have come to realize that the natural habitat is important for people too. For too many years, the wetlands and the Bay have been treated like dumps, repositories for whatever waste that the residents didn't want. Wetlands were filled and developed. Rivers were channeled and controlled. Now efforts are underway to restore wetlands and other wild areas for the benefit of plants and animals, but also for humans.

Of course, there is only so much that humans can do to restore or even maintain the region as it is. The Bay was not always here, but the geological forces that created it are still active and will one day destroy the Bay. Here we will attempt to describe the Bay in the past, present, and future, and to explain those forces. The most powerful are the geologic forces, and indeed, several significant faults cross the Bay and surrounding area, and thus, it is and has been subjected to considerable tectonic activity over the years. Earthquakes and ground movement are the major forces that built the San Diego Bay region, and periodic earthquakes always will rock the region. The largest most recent earthquake was a magnitude 6.0 in 1862. There is an 18% chance of a 6.7 or greater magnitude earthquake occurring in San Diego County in the next 30 years. Movement on those faults is what built the region and what will eventually destroy it. Other forces include the rise and fall of the ocean with changes in climate, atmospheric rivers, and finally humans. In the crush of our daily activities, it's easy to forget that these forces are not influenced by our meetings, jobs, smartphones, or desires. They move at their own speed and come in their own time. We are now feeling the effects of the climate crisis with much hotter temperatures and seemingly endless fire seasons. The sea level is beginning to rise as the Earth's large icecaps and glaciers melt. That will change the coast and the region over the next years. Earthquakes periodically rumble through the region to remind us that we are still part of the natural world.

This book will describe the natural history and evolution of the San Diego Bay region (e.g., its geology, plants and animals, people and their activities, and the connections among these) over the last 50 million years through the present and into the future. We will explain how those forces that built the Bay are still active today. Scientists have already made predictions of what might happen in the future. We will review the predictions about the future of the Bay region and how those forces will eventually change the region beyond recognition. Humans witnessed the end of the Ice Age, the rise of sea levels, and the flooding of the Bay. We do not know if humans will see the end of the Bay, but we do know that change is coming.

2 Geological Forces that Built San Diego Bay

The story of the San Diego Bay Area begins deep in the Pacific Ocean about 165 million years ago, roughly in the early Late Jurassic. The supercontinent Pangaea had just broken into two smaller supercontinents, and the Atlantic Ocean was forming in the gap between them. The continents were ruled by dinosaurs, such as the *Brontosaurus*, *Stegosaurus,* and *Allosaurus,* the largest predator of all at the time. Most of California did not exist. The Pacific Ocean reached all the way east to Arizona. Over the succeeding millions of years, enormous geologic forces created Southern California and the San Diego Bay and area. Those forces include plate tectonics, volcanic activity, crustal uplift and depression, folding, and erosion. The result is a complex jumble. Interestingly, those forces are still active today.

The landmass that would later become the San Diego Bay region resulted from titanic forces that have and continue to shape the continents. To all of us, the ground we stand on seems solid. In fact, it isn't. It is in constant motion and ever changing. The energy comes from the Earth's superhot core. The inner core is a solid mass of mostly iron at 7600 to 13,000°F. The outer core is molten iron and nickel at 5800 to 9400°F. Between the core and the crust is the mantle, which is also extremely hot and continually moving. The crust makes up only a relatively thin layer on top of the mantle (about 18 miles thick on the continents and about 3 miles under the oceans). It is divided into massive plates that "float" on the mantle far below the surface. The movement of the plates is powered by the convection in the mantle. This forces new material up through weak areas in the crust, such as the spreading areas or rifts in the middle of the oceans. New material upwells from the mantle, is pushed along over the surface, and is later subducted where plates meet. Along those edges, earthquakes and volcanoes are common. Ultimately, the energy for the movements comes from the Earth's cooling, but still, fantastically hot mantle and core.

BUILDING SOUTHERN CALIFORNIA

The forces noted above all combined to extend the western coast of North America from Arizona and Utah to the current coast of California. Much of California is a jumble of materials from multiple sources. Some came from the floor of the Pacific Ocean. For example, the deposits of uncountable numbers of microscopic radiolarians formed chert. Other calcareous organisms formed layers of sediment that turned to limestone, and sand transformed into sandstone. These sedimentary rocks were scrapped up onto the growing coast. Others fell into the ocean from the granite mountains of the original west coast and were

DOI: 10.1201/9780429487460-2

pushed back onto shore. Magma columns poked holes in the other rock and poured out additional material. The tectonic and other forces then continued to compress and stretch the land into mountains and valleys. In the end, Southern California and the San Diego Bay area developed into what we see today.

We typically associate earthquakes with California. However, the formation of the southern part of the state did not initially involve faults and earthquakes. Those came later. In fact, the earliest beginnings of California were relatively quiet. The landscape along the Arizona coast 200 million years ago was low and rolling with rivers that no longer exist (Walawender, 1999). Those rivers eroded the granite rocks that made up the mainland, and the sediment flowed down the rivers and into the ocean to form an alluvial fan. Much of the geologic record of those years and later is contained in the rocks east of San Diego. One of those formations is the Julian Schist, which includes those ancient sediments that washed down the rivers and into the shallow seas just off the coast at that time. After many years of sediments piling up, they were transformed into mudstone and sandstone. Later, columns of magma penetrated the sediments and distorted their horizontal layers into near vertical plates. Even further east, marble deposits represent the remains of limestone sediments laid down in shallow inland seas.

Plate Movement

Plate tectonics was the primary force that formed California. Movements of the tectonic plates are caused by movements in the mantel due to the heat there. The interactions at the boundaries of the plates are of three types: convergent, divergent, and transform. The convergence can be oceanic-oceanic, continental-continental, or oceanic-continental. In oceanic-oceanic convergences, one or both of the plates will override the other and be lost by subduction. In continental-continental, they run directly into each other to produce a crumpled boundary that creates large mountain ranges, such as the Himalayas. In oceanic-continental, the denser oceanic is subducted under the lighter continental crust. Divergent plates move away from each other. In transform boundaries, the two plates slip along against each other. The creation of California first involved subduction and later a transform boundary.

The continental plates have alternately existed as a huge single supercontinent or multiple smaller continents. The supercontinents generally last about 100 million years before they break up and the resulting continents drift apart. Every few hundred million years, the cycle repeats itself. At least three supercontinents have existed over the past 2 billion years. There is some evidence for others, but it is not compelling. Nuna is the oldest known supercontinent known so far. It was in place about 1.8 billion years ago. The next was Rodinia about 1 billion years ago, and finally, Pangaea was about 300 million years ago. At the other times between those three, smaller continents, such as we have today, existed. Scientists use the magnetization of iron to show how those continents interacted in a supercontinent. In molten rock, magnetized iron oriented itself along the magnetic lines of the Earth. Once the rock solidified, its history was locked in stone and can be compared to the magnetic orientation of other rocks. By comparing iron-containing rocks from different regions and knowing the time frame, scientists can put the puzzle back together again and derived the movement of the continents and their joining to form a supercontinent.

The California coast developed as a result of the movement of the massive pieces of the Earth's lithosphere, which includes the crust and upper mantle. In the cracks between the plates (e.g., mid-ocean rifts) and in other areas, hot material from the mantle rises to the surface and pours out to form a new crust. As the new material is forced to the surface at mid-ocean rift zones, it pushes the plates away from the rift in different directions. The new rock begins as pillow basalt that erupted thousands of miles out in the Pacific Ocean at the mid-ocean rift 100–200 million years ago. That material entered a "conveyor belt" that moved it from its origin to the coast over millions of years (Figure 2.1).

Subduction

That conveyor belt pushed the massive Farallon plate eastward, where it collided with the North American plate. Since oceanic plates are denser than continental plates, the Farallon plate sank below the North American plate in a process

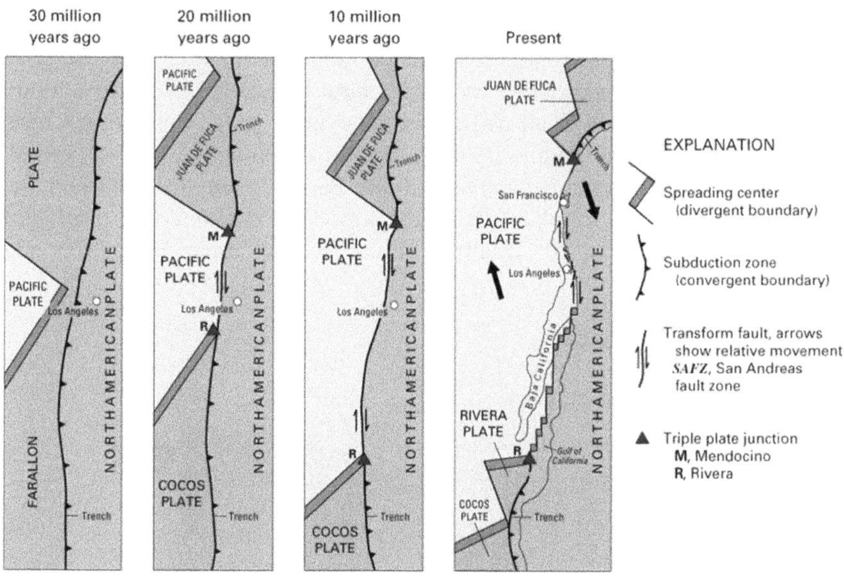

FIGURE 2.1 Farallon plate. The making of the San Diego Bay region involved the enormous forces that move the Earth's tectonic plates around the surface. Spreading of the mid-oceanic ridge pushed the Farallon plate toward the east so that it collided with the North American plate, which was moving west. The Farallon plate was subducted under the North American plate (Diagram courtesy of US Geologic Survey).

called subduction (Figure 2.2). During the subduction, some material was scrapped up from the ocean bottom and deposited on the edge of the North American continent. This added to the growing area that would become California. Subduction also creates enormous amounts of heat as the plates grind past one another deep underground. That heat can yield volcanic activity at some distance inland from the plate boundaries. Also at the boundary, large pieces of the eroded material and the rock called terranes broke off from the North American plate and fell into the ocean floor just off the then-coast. As the subduction continued, those pieces were collected and became attached to the two plates. This was particularly true in Southern California. The Farallon plate is now completely subducted under the North American plate. A few pieces of that plate broke off (e.g., Juan de Fuca plate off Oregon and Washington).

The material that formed Southern California came from several sources. First, the hard granite and metamorphic rocks at the edge of the continent were slowly eroded into layers of sediment. Over the years, the sediments collected in the shallow waters just off the then coast of Arizona and Nevada. They eventually solidified into sandstone and mudstone. Some of the rocks in the complex were subjected to tremendous amounts of heat and pressure. These rocks changed or metamorphosized. All of these rocks formed layers on one another, and then those blocks were broken and reconfigured as the plates crashed into each other. Second, material from the ocean floor was scrapped up and added to the growing front of California. That included chert that formed from the countless numbers of microorganisms called radiolarians. Once they die, their silica shells fall to the ocean floor and harden into chert. Some of the deposits are 250 feet thick. Third, in Southern California, a small plate, which might have been a fragment of the Farallon plate, added to the mixture as it was caught between the North American plate and the Pacific plate that followed the Farallon plate. That small plate affixed itself to the North American plate until a part of it was subducted. Finally, volcanic activity broke through to add

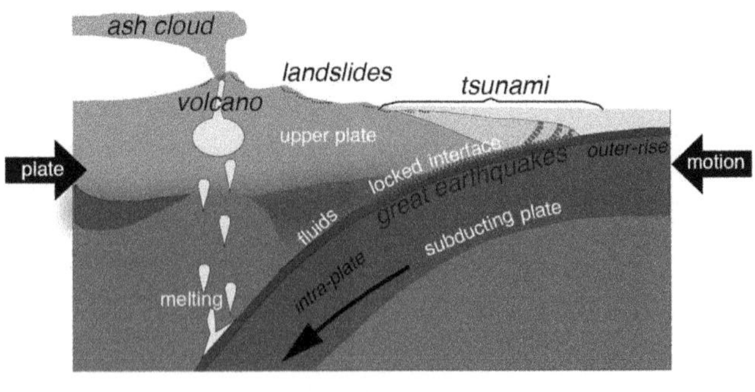

FIGURE 2.2 Subduction. Over many years, the heavier oceanic Farallon plate was subducted under the lighter continental North American plate. Most of the Farallon plate is now underneath the North American plate. (Illustration from US Geologic Survey).

more material. The friction of the Farallon and North American plates moving against each other caused the subducted material to melt. The resulting plumes of melted rock caused volcanic activity that formed the volcanoes in present-day Eastern California. The Farallon plate has disappeared entirely underneath the North American plate. Yet, its influence has been felt several times over the ages. For example, its movement and the resulting heat were responsible for the massive earthquakes on the New Madrid fault in the Mississippi River Valley in the early 19th century (Forte et al., 2007). In addition, subduction of a fragment of the Farallon plate, the Juan de Fuca plate, is providing the energy for the Cascadian volcanoes.

Once the Farallon plate was completely subducted, the Pacific plate collided with the North American plate, and a new transform boundary formed between the two plates. It is now a strike-slip boundary in which the two plates grate past each other. The new boundary resulted in the San Andreas fault system. The exact location of the boundary is not always completely clear. It moves, depending on the stresses involved in the movement of the plates against each other. When the plates are roughly in a straight line, the movement is fairly slow and steady. Of course, it is not always a straight line, and those twists and turns put stress in the system that has to be relieved in some way. For example, the area behind a bend becomes compressed and pushed up. This is termed transpression. The Rose Canyon curves to the west in La Jolla, and the resulting transpression yielded Mount Soledad (Dewey et al., 1998). In fact, Mt. Soledad is still slowly gaining height. In other areas, the landmass is stretched and becomes thinner. This is called transtension. Transtension causes the land to droop to form the valleys and basins. San Diego Bay is such a depression. San Diego Bay, Mission Bay, and other areas are bounded by the Rose Canyon and Point Loma faults. Depressions, such as this, are called a nested graben. The movement apart of two faults results in a depression or graben between them. In cases of transpression and transtension, the stress is relieved in the third dimension. In transpression, the ground rises, and in transtension, it sinks. The fault continues to move. Some sections creep as they have for some time, and they tend to have multiple small earthquakes, and those earthquakes can be used to characterize the faults. Those creeping faults are less likely to build up large amounts of energy and release it suddenly when the fault ruptures, resulting in a large earthquake. About 3.5 million years ago, the nature of the interaction of the Pacific and North American plates changed. They began to partially oppose one another rather than simply slide along each other.

Interestingly, during these tectonic activities, the Pacific plate managed to capture part of the North American plate. All of the land to the west of the San Andreas fault, including much of the California coast, was transferred to the Pacific plate and is now moving northwest with that plate as the two plates slip past one another (Meldahl, 2015). In the future, that part of California and possibly more will continue to migrate with the Pacific plate (see Chapter 10).

Buoyant density was also involved in the movement of the crust. The lithosphere includes the crust and top part of the mantle. The mantle just under the lithosphere is the asthenosphere. It is a semi-solid but flows. The relatively cold continental mantle lithosphere is denser than the asthenosphere below it. This action causes the lithosphere to thin and the hotter asthenosphere to move upwards. That inversion is called delamination. For example, delamination occurred as the Farallon plate was subducted under the North American plate (Putirka and Busby, 2011), and the uplift resulted in the Sierra Nevadas and the volcanic activity in eastern California.

Peninsular Batholith

The Peninsular Ranges are part of a large batholith that at one time stretched from Northern California to Southern Baja California (Silver and Chappell, 1988; Todd et al., 2003). Batholiths are large masses of igneous rock that form from magma that cools below the surface. Plutons and batholiths are related terms. Both are the remnants of magma that solidified underground. Plutons are the smaller version. They may represent the magma chamber of a single volcano. A batholith contains many plutons that resulted from multiple intrusions. These are common along subduction zones. That massive amount of igneous rock of the Peninsular Ranges was deposited in one sequence of magmatic activity that created two batholiths. The western one formed 105–150 million years ago in the early Cretaceous by subduction. The eastern one formed 80–105 million years ago.

The two plutonic belts are products of the subduction process and are partial melts of very large amounts of material. The two in Southern California came from two different events and places. The western belt seems to have originated about 110 million years ago and about 1000 km south of San Diego in the Pacific Ocean. The belt formed as a long narrow arc, and over time, the small plate carrying it moved to its present position and collided with the growing coast of California. This arc now stretches from Riverside through Baja California. The western belt is unusual (Langenheim and Jachens, 2003). It seems to be a solid block between the two huge plates, and it has maintained its integrity for some time. These factors may also have implications for the rifting in the Gulf of California.

Interestingly, the small plate carrying the western belt was abutted to the North American plate for millions of years, and then, part of it broke loose to be subducted under the western belt and the much large plate. While many volcanoes result from the friction of one plate becoming subducted under another, not all do. Sometimes a subducted plate develops holes that will allow magma to well up (McCrory et al., 2009; Hawley and Allen, 2019). The subduction allowed the asthenosphere beneath the fragment to upwell and melt more material and that large melt filled the gap between the western belt and the North American plate to form the eastern belt (Hildebrand and Whalen, 2014).

Jiang et al. (2019) examined the two batholiths. They took a transect of southern San Diego County and characterized all of the plutons so that they could be assigned to ten plutonic suites. The western group included five granitic and one gabbroic plutons. There were produced by a Cretaceous intrusion. The eastern group is younger. The two batholiths differ in chemical composition. The western has more mafic rocks, which are silicate mineral or igneous rock that is relatively rich in magnesium and iron. Mafic rocks tend to be dark in color (e.g., olivine, pyroxene, amphibole, and biotite). The eastern includes more felsic rocks, which are igneous rocks that contain feldspar and quartz. The surface is characterized by uplifts and depressions at the faults (Meldahl, 2015). Rapidly uplifting mountains end in large triangular facets at the end of the ridge at the fault and small alluvial fans. Slowly uplifting mountains have small triangular facets and large fans.

MAJOR FAULTS

The North American and Pacific plates continue to slip along past each other in fits and starts. For us, the "fits" are long periods of stasis when everything seems solid. The "starts" are earthquakes that release massive stresses that have built up in the land for those times that the slippage has been hindered. The San Diego Bay region has fewer earthquakes than the San Francisco Bay Area. However, a number of faults near San Diego are still capable of causing an earthquake of magnitude 7.0 or more. As in Northern California, the San Andreas in Southern California is a system of faults that account for movement along the transform boundary between the Pacific and North American plates as those two plates slip along past each other.

San Andreas Fault

The San Andreas fault is the largest and best-known of the faults along the California coast. The fault lies east of San Diego, and so, most of the San Diego Bay and surrounding areas are on the Pacific plate. The fault ends near Bombay Beach in the Salton Sea, but an extension, called the Imperial fault, continues south to the Cerro Prieto spreading center. Cerro Prieto is a volcano that lies astride the East Pacific Rise system that is slowly rifting Baja California away from mainland Mexico.

The San Andreas Fault is about 100 miles to the east of San Diego, and so, it is not in the San Diego region. Nevertheless, it is an important fault for the region and has an influence on the city and region. First, earthquakes on the southern end of the fault can sometimes be felt in the city. Second, the geologic future of the city is tied to the fault as is explained in more detail in Chapter 10.

The fault was discovered by Andrew Lawson (University of California, Berkeley) in 1895. It formed about 30 million years ago and is the boundary between the Pacific plate and the North American plate. The Pacific plate is moving roughly northwest, and the North American plate is moving westward. The motion along the fault is technically called a right-lateral strike-slip fault. The fault is actually part of a system of faults that run 750 miles from the Salton Sea in Southern California to the Mendicino triple junction in Northern California. Earthquakes occur periodically on all of the faults as the stress from the movement of the plates builds up. After it reaches some threshold, the energy is released as the plates slip past each other to a new position. Most earthquakes are very small, but some quite large earthquakes have taken place in the last 100 or so years, and the potential for another large earthquake remains.

The movement on the San Andreas fault can readily be seen. One of the most impressive examples is the Pinnacles, which are located today just east of Soledad. When the Neenach volcano erupted 23 million years ago, it was near the present-day city of Lancaster nearly 200 miles south of its current position. The volcano straddles the fault. As the Pacific plate has continued to move northward, half of the volcano moved with it. The other half remained in place in Southern California. The distinguishing feature of the Pinnacles is the rock columns that are so attractive now to rock climbers. Those columns represent the magma that remained in the volcanic chimneys and solidified before it could erupt. Over millions of years, the softer rock that surrounded the columns has eroded away, leaving the harder columns standing alone. In addition, the main trace of the San Andreas fault has move 4 miles to the east. Over the last 12 million years, the San Andreas fault has moved 175 km (Langenheim et al., 2010).

San Jacinto Fault

This fault lies just west of and roughly parallel to the San Andreas fault. Sanders (1993) noted that earthquakes on the two faults seemed to be related. For example, the average recurrence intervals are similar (San Jacinto: 150 years; San Andreas: 132 years). Also, the large 1857 El Tejon earthquake triggered earthquakes on the San Jacinto fault. Large earthquakes on this fault in 1899 and 1918 resulted in significant damage to surrounding towns.

Elsinore Fault

This fault is just west of the San Jacinto fault. At 110 miles, it is the longest fault in the area, and it includes several others, such as the Whittier, Chino, and Laguna Salada faults. Finally, it is the quietest of the faults. The last major earthquake was a magnitude 6 event in 1910 near the city of Lake Elsinore. There is some evidence that it is linked to other faults in Mexico.

Rose Canyon Fault

The Rose Canyon fault is important because it runs for about 19 miles through the most populated areas of San Diego (Figure 2.3). It begins in Mission Valley and runs up to La Jolla and into the Pacific Ocean. This fault is the most seismically active fault system in Southern California (Ross et al., 2017). In the last 120 years, 11 earthquakes (magnitude > 6) have occurred there. For part of its trace near Borrego Springs, it splits into three smaller faults (Coyote Creek, Clark, and Buck Ridge) referred to as the trifurcation area. It is potentially more dangerous because the Rose Canyon and Newport-

FIGURE 2.3 Rose Canyon and other major faults. The Rose Canyon fault is the most dangerous in the San Diego Bay region because so much of its trace is through populated areas. Several other faults lie to the east, including the Elsinore and San Jacinto faults. Even further east is the San Andreas fault. In addition, there are faults just offshore, such as the Point Loma and Coronado Bank faults (Map courtesy of US Geologic Survey).

Inglewood faults are continuous. They form a 170-km-long fault system that is capable of severe earthquakes.

The last major earthquake on this fault was in 1933, but it was north of the San Diego Bay Area in Long Beach. The last earthquake on the fault in the San Diego Area was thought to be in 1862 and was estimated to be about a magnitude 6. Lindvall and Rockwell (1995) found that the fault has had at least four surface rupturing earthquakes, in the past 11,000 years and accumulated at least 8.7 meters of lateral displacement in less than the past 8100 years. Rockwell et al. (2018) studied two paleoseismic trenches at Old Town in San Diego and found that the area had experienced six earthquakes in the past ~3,300 years. Three significant events that were more severe than the 1862 event (Singleton et al., 2019). Those three took placed between 3300 and 700 years ago.

Point Loma Fault

This fault runs roughly parallel to the Rose Canyon fault. The region between this fault and the Rose Canyon fault forms a nested graben or depression that is now the San Diego Bay.

Others

There are also faults offshore. The Coronado Bank Fault Zone is about 12 miles offshore and essentially continuous with the Palos Verdes fault. The San Diego Trough fault is a bit farther offshore. The San Clemente fault is farther still offshore. The most recent surface ruptures of all three were in the Holocene.

VISIBLE REMINDERS OF THE FORCES THAT BUILT THE BAY

Although the geologic forces that made the San Diego Bay Area are still just as active, they move at a rate that is far longer than the ability of human observation to perceive. Those geologic forces required millions of years to bring it to what it is today. The Farallon plate was pushed into the North American plate by new material forcing it from the mid-ocean rift. The heavy Farallon oceanic plate was subjected underneath the lighter North American continental plate. The remnants of the Farallon plate broke off to allow windows of hot magma to flow upwards and add to the sedimentary rocks from the two plates. The extreme slow movement of geology is nothing but an illusion. With the exception of the occasional earthquake, we cannot perceive their activity. However, those forces are not weaker or slower than they ever were, and the evidence of the geological past can easily be seen in the San Diego Bay Area today.

Mount Soledad

This mountain formed by transpression of the land at a point where the northernly Rose Canyon fault makes a west turn (Figure 2.4). The left turn has great significance. The Rose Canyon fault is a strike-slip fault. Each side of the fault slips along as long as the fault forms a straight line. The left turn represents a binding region where stresses can increase. So the land is piling up at this point, and it has not where to go but up. That action pushed the Cretaceous-era seabed and Eocene-era river deposits up 823 feet to form Mount Soledad. Those movements also resulted in La Jolla Cove and the depression or graben below sea level that created San Diego Bay.

From the top of the mountain, other geologic features can be seen. To the northwest just south of piers at Scripps Institute of Oceanography is the La Jolla Canyon. Of course, most of that is underwater and invisible. To the north and northeast are marine terranes that appear as steps. The older terranes are to the east, and the canyons were carved into the terranes when they were at sea level. The oldest of the mesas were in the water 1.3 million years ago.

La Jolla Cave and the Coast Walk Bridge

The Rose Canyon fault is actually a fault system with several smaller segments. One of those segments can be seen at the surface in La Jolla. Just under the Coast Walk Bridge near the Matlahuayl State Marine Reserve, the Country Club fault

FIGURE 2.4 Mount Soledad. Mount Soledad resulted from a bend in the Rose Canyon fault. Movement along the fault is hindered by the bend, resulting in transpression and uplift. In fact, Mount Soledad continues to gain height every year (Photograph reproduced with permission from Sebastian Kaser).

goes offshore for about 84 miles until it reaches Newport Beach (Figure 2.5). Thus, La Jolla Cove is moving northward, and La Jolla Shores and University of California, San Diego are moving southward.

The bridge crosses the exact point at which the County Club segment of the Rose Canyon fault goes off land and into the ocean. Everything to the west of the fault is on the Pacific Plate and moving northwest. Everything to the east is on the North American Plate and moving south. The Rose Canyon fault system is the most dangerous fault in the San Diego area. Two more segments, the Mount Soledad and Rose Canyon segments are north of the County Club segment. The slip rate on the fault is about 1.1 mm/year. The whole Rose Canyon-Newport-Inglewood fault system is thought to be capable of an earthquake of 7.0 or greater, and because it runs through a highly populated area, it is considered to be a serious threat to the region.

Temecula Gorge

This region began to form about 225 million years ago under the Pacific Ocean (Chester, 2001). Sediments collected on the ocean floor until about 200 million

FIGURE 2.5 Coast Walk Bridge in La Jolla. Below the bridge is the exact site where the Country Club fault goes offshore. It is one component of the Rose Canyon fault system. Two other segments run offshore north of this site (Photograph reproduced with permission from Sebastian Kaser).

years ago, when they were subducted under North American plate and hardened into sandstone and other material that became the Bedford Canyon Formation. Then about 150 million years ago, more oceanic crust subducted under the Bedford Canyon Formation. Heat from the subduction melted enormous amounts of rock. Some of the melted rock made it to the surface in a chain of volcanoes. Over about 74 million years, the heating and earth movements reshaped the Bedford Canyon Formation so that it is no long horizontal layers of sedimentary rock. One result is Temecula Canyon, where those layers are now vertical. Today Temecula Canyon shows several remnants of its geologic history. The reddish rocks on the north side of the canyon are the Bedford Canyon Formation. On the south side, are the Woodson Mountain granodiorites, which are the beige granitic remains of the volcanoes.

Calavera Hills

Calavera Hills has a dormant volcano just north of San Diego. The volcano is assumed to have formed in the Miocene. The volcano broke through other older formations in the area. This volcano was not caused by the earlier subduction. As the new boundary between the Pacific and North American plates began to form, the Pacific plate suffered fractures and faulting. fracturing and faulting of the Pacific plate as the boundary between the two plates was forming. As the remains of the warmer, lighter spreading center between the plates was subducted under the North American plate, the crust lifted up and fractured to create weaknesses where magma could break through, resulting in these volcanos. When Mount Calavera last erupted about 22 million years ago, magma cooled into geometric columns before it could be spewed out. A plug of fine-grain basalt formed in the volcano and blocked it. The plug can be seen today in the gray rock in the walls of the ancient caldera. The rock is dacite, which is somewhere between a rhyolite and an andesite in composition. This igneous rock cooled slowly underground and was later exposed by erosion and used to produce gravel. Overlying that formation is the Santiago formation, which is soft sedimentary rock (e.g., sandstone, siltstone, and claystone) deposited in the Eocene.

Black's Canyon

A site near the beach features maroon-purple pebbles that are the remnants of a volcano in Sonora, Mexico. These pebbles have moved north to the San Diego region as the Pacific Plate moves steadily northward (Meldahl, 2015).

Point Loma

Point Loma formed about 80 million years ago as sediment flowed from the eastern mountains to form a fan in what the Pacific Ocean of that time (NPS, 2020). Sediment accumulated for millions of years and solidified. Later compression on the Rose Canyon fault uplifted the peninsula. It now contains some very old rock formations. The Point Loma Formation on the west side formed in the Late Cretaceous period about 75 million years ago (Figure 2.6). The rocks contain dinosaur and other fossils. The Cabrillo Formation lies on top of the Point Loma Formation and was also formed in the Late Cretaceous.

Los Peñasquitos Canyon

The Peñasquitos Formation is the oldest rocks in the canyon. They are best seen at the waterfall. They formed 140–150 million years ago. Volcanic ash, clays, silts, sands, and sandy pebbles collected along the then continental edge. These were subjected to heat and pressure and transformed into shales and metavolcanic rocks when lava form the Southern California Batholith intruded.

San Diego Formation

This conglomerate and sandstone deposit underlies a large area from Mount Soledad to northern Baja California. It contains the San Diego Formation Basin, which is a large aquifer under Imperial Beach, Chula Vista, National City, and

FIGURE 2.6 Point Loma Formation. These sedimentary rocks are among the oldest in the region. They formed from the compression of eroded material about 80 million years ago (Photograph is in the public domain).

the southern part of San Diego. It contains 960,000 acre-feet of brackish water. The Richard A. Reynolds Groundwater Desalination Facility in Chula Vista draws water from the aquifer to provide additional diversity to the sources of water for the region.

Julian Schist

These formations are among the oldest rocks in the San Diego Area. When the west coast was near Arizona, sediments from the North American continent eroded into an alluvial fan in a shallow sea. Over time, those sediments piled up, and the pressure of additional material on top fused the original sediments into rock. Outcrops can be found throughout the eastern part of San Diego County, but some easily visible areas are near Mount Laguna on the Sunrise Highway.

CONCLUSION

Multiple geologic forces have combined to build the San Diego Bay region. Plate tectonics, subduction, delamination, plutons, transpression, transtension, volcanism, erosion, sedimentation, earthquakes, and many more processes were involved. All of these forces also had millions of years to work on the Earth's crust in this relatively small area. Interestingly, these forces are all still active today and will eventually change the San Diego Bay region beyond recognition (see chapter 10).

REFERENCES

Chester T (2001) Geology of Temecula Canyon. Retrieved from: http://tchester.org/fb/geology/temecula_canyon.html; accessed February 21, 2021.

Dewey, JR, JE Holdsworth RE, Strachan RA (1998). Transpression and transtension zones. In: *Continental Transpression and Transtensional Tectonics* (eds. Dewey, JR, JE Holdsworth RE, Strachan RA) Geological Society, London, Special Publications, 135: 1–14.

Forte AM, Mitrovica JX, Moucha R, Simmons NA, Grand SP (2007) Descent of the ancient Farallon slab drives localized mantle flow below the New Madrid seismic zone. *Geophysical Research Letters* 34: L04308.

Hawley WB, Allen RM, (2019) The fragmented death of the Farallon plate. *Geophysical Research Letters* 46: 7386–7394.

Hildebrand RS, Whalen JB (2014) Arc and slab-failure Magmatism in Cordilleran Batholiths II – The Cretaceous Peninsular Ranges Batholith of Southern and Baja California. *Geoscience Canada* 41: 399–458.

Jiang H, Lee CTA (2019) On the role of chemical weathering of continental arcs in long-term climate regulation: A case study of the Peninsular Ranges batholith, California (USA). *Earth and Planetary Science Letters* 525: 115733.

Langenheim VE, Jachens RC (2003) Crustal structure of the Peninsular Ranges batholith from magnetic data: Implications for Gulf of California rifting. *Geophysical Research Letters* 30: 1597.

Langenheim VE, Graymer RW, Jachens RC, McLauglin RJ, Wagner DL, Sweetkind DS (2010) Geophysical framework of the northern San Francisco Bay region, California. *Geosphere* 6: 594–620.

Lindvall S, Rockwell TK (1995) Holocene activity of the Rose Canyon fault zone in San Diego, California. *Journal of Geophysical Research* 100: 24121–24132.

McCrory PA, Wilson DS, Stanley RG (2009) Continuing evolution of the Pacific–Juan de Fuca–North America slab window system—A trench–ridge–transform example from the Pacific Rim. *Tectonophysics* 464: 30–42.

Meldahl KH (2015) *Surf, Sand, and Stone*. University of California Press, Oakland, CA.

NPS (2020) Geology of Cabrillo National Monument. National Park Service. Retrieved from https://www.nps.gov/cabr/learn/nature/geology. htm#:~:text=The%20elongated%20mass%20of%20eroded,sediment%20called%20a%20submarine%20fan; accessed February 10, 2021.

Putirka K, Busby C (2011) Introduction—origin and evolution of the Sierra Nevada and Walker Lane. *Geosphere* 7: 1269–1272.

Rockwell T, Singleton D, Murbach M, Murbach D (2018) Mid to late Holocene rupture history of the Rose Canyon Fault in San Diego, California. US Geologic Survey. Retrieved from https://earthquake.usgs.gov/cfusion/external_grants/reports/G16AP00015.pdf; accessed February 10, 2021.

Silver LT, Chappell BW (1988) The Peninsular Ranges batholith: An insight into the evolution of the Cordilleran batholiths of southwestern North America. *Earth and Environmental Science Transactions of the Royal Society of Edinburgh* 79: 105–121.

Singleton DM, Rockwell TK, Murbach D, Murbach M, Maloney JM, Freeman T, Levy Y (2019) Late-Holocene rupture history of the Rose Canyon fault in Old Town, San Diego: Implications for cascading earthquakes on the Newport–Inglewood–Rose Canyon fault system. *Bulletin of the Seismological Society of America* 109: 855–874.

Todd V, Shaw SE, Hammarstrom JM (2003) Cretaceous plutons of the Peninsular Ranges Batholith, San Diego and westernmost Imperial Counties, California: Intrusion across a Late Jurassic continental margin. In (eds. Johnson, S.E., et al.) *Tectonic Evolution of Northwestern Mexico and the Southwestern USA*: Geological Society of America Special Paper 374: 185–235.

Walawender MJ (1999) *The Peninsular Ranges: A Geological Guide to San Diego's Back Country*. Kendall Hunt Publishing Co., Dubuque, IA.

3 Water

Water is a powerful force, especially when it has millions of years to work. The water that feeds the San Diego Bay and surrounding areas comes from several sources, including precipitation, river flow, residential and industrial releases, and of course, the Pacific Ocean.

OCEAN WATER

Pacific Ocean

San Diego and Mission Bays both open to the Pacific Ocean, and most of the rivers in the area flow directly or indirectly into the Ocean. The Bays are tidal, and the water in the Bays is heavily influenced by the Ocean. However, the coast of California has two segments. On the northern segment, the coast is oriented in a roughly north to south manner. A person standing on the coast and looking out to sea is facing west. However, the direction of the coast changes dramatically at about Point Conception. From there on south, a person standing on the coast is facing south. This broad curve of coast is known as a bight. In this case, it is the Southern California Bight. The ocean currents and waves are affected differently by these two different orientations.

The Bight is in the middle of several different water masses, including the Pacific subarctic, Pacific equatorial, and the North Pacific central water masses. The water circulation patterns of the Bight are more complex than other points on the west coast. The California current system has three components (Bray et al., 1999). The current itself generally consists of slow cold water flowing south from the North Pacific. Another current mostly along the continental slope flows northward. Finally, in the Southern California Eddy, the California current turns to the north within the Bight. In the winter, the surface flow is toward the north. In the spring, the winds turn the flow south. The movement of the water can be involved in shoreline and cliff erosion and the distribution of run-off materials.

During the late Neogene and the late Miocene and Pleistocene, conditions in the Bight allowed an amazing amount of marine diversification (Jacobs et al., 2004). These involved the estuaries of the mid-California coast and the opening of the Gulf of California, but the result is a much more diverse collection of marine species than are found on the East Coast.

Being located on the coast, San Diego Bay and the surrounding region are at some degree of risk for tsunamis. Barberopoulou et al. (2011) examined that the risk for local and distant tsunamis. They showed that for the most part, the

DOI: 10.1201/9780429487460-3

largest waves in the San Diego Bay are from local events. Only the Alaska-Aleutian Trench offers a distant risk. This is mostly because the Bay is so well protected by North Island and the Silver Strand. However, large currents can occur inside the Bay. A massive earthquake of magnitude 8.8 in Chile provided a real-world test of their findings, and indeed, the predictions were validated.

Sea Level Rise

In the late Pliocene about 2.67 million years ago during the last Ice Age, much of North America was covered with vast ice sheets. In fact, about a quarter of the Earth's continents were covered in ice. Later as the ice melted, the sea levels rose so that Pacific Beach, Tijuana, Coronado, North Island, and much of Point Loma were underwater. The water was more than 60 meters above the current levels. This back and forth of San Diego underwater or dry land has played out multiple times as the ice sheets come and go. For next several hundred thousand years, sea levels rose and fell as the Ice Ages cycled. About 15,000 years ago, the last cold period ended, and the ice began to melt. Sea levels rose some 90–120 meters.

Sea levels have risen and fallen many times as the Earth experienced Ice Ages in which massive amounts of water were frozen in the ice caps and glaciers that covered much of the Northern Hemisphere. As the Earth moved from Ice Age to Ice Age, sea levels rose and fell, and the Bay filled and emptied. The last of the Ice Ages occurred about 30–40,000 years ago. During that last Ice Age, San Diego and Mission Bays were dry plains with hills and a couple of rivers that ran across a plain some 90 km to finally reach the Pacific Ocean. About 20,000 years ago, sea levels dropped about 120 meters due to the Ice Age. As a result, Southern California had few estuaries (Dolby et al., 2016). Estuaries form when the slope at the shoreline that separates fresh and seawater is gentle and shallow. The loss of so much ocean water brought that interface away from the gently sloping continental shelf to the steeper continental slope. Dolby et al. found evidence for only two. One was near Morro Bay, and the other was halfway done Baja California. In other words, much of the California shore was devoid of estuaries until the ice melted and sea levels rose again. As the Ice Age ended 8–12,000 years ago, about 10,000 years ago, sea levels rose, and water eventually flooded the Bay as we know it today. That pattern of a wet and dry Bay has been repeated multiple times over the ages.

Evidence of the rise and fall of sea levels can be seen around the San Diego area. Near the coast, the ancient shoreline is represented by raised marine terraces and cliffs. Near La Jolla, at least three terraces are found at the edge of Mount Soledad. In the Clairemont Mesa area, a ridge near Convoy Street is a part of an ancient beach ridge. Others can be seen in the area. In 1977, Atwater et al. reported two sites of estuarine deposits that showed at least two times that the water in the Bay was substantially lower than it is today. They were the Sangamon and post-Wisconsin high stands of sea level (see Otvos, 2014, for a description of those times). When the water came back into the Bay 10,000 years ago, it entered at about 2 cm per year and spread about 30 meter per year until about 8,000 years ago when it slowed down to the present rate of about 0.1–0.2 cm per year.

FRESH WATER

Precipitation

San Diego averages less than 30 cm of rain per year, and most of the rain falls from December through March. Rain is rare in the summer. Hurricanes are also rare, but not unknown (Chenoweth and Landsea, 2004). The wettest year on record was 1941 with just under 90 cm. The driest was 1953 with only just over 8 cm. The demand for water in San Diego and California overall has grown with the burgeoning population, and large water projects have been used to attempt to equitably manage the water for people, nature, and agriculture. Most rivers were dammed. Water from Northern California is even routed as far south as San Diego.

Overall, California has typically experienced periods of more or less precipitation. In any year, two-thirds of the state's precipitation falls in Northern California. The cycles have covered decades of time. For the last 50 years, the state has been in a relative wet phase. The Western United States has experienced extreme fluctuations in rainfall. These extremes can easily be demonstrated by the 2012–2015 drought in California that was followed by winter storms that resulted in floods and mudslides in 2015–2016 (Barth et al., 2016). Barth et al. also noted that more than 80% of these severe weather events are associated with atmospheric rivers in Northern California.

Periodically, California has been struck by atmospheric rivers. These are very large-scale precipitation events that involve 100–150 mm of rain in a 24-hour period seem to occur about every 2 years (Cordeira et al., 2019). However, those numbers can be much greater, up to 600 mm in some cases. In the winter of 1861–62, 90 cm of rain fell in 30 days. The Sacramento and San Joaquin valleys were flooded, and more than 3 million acres were underwater. Most of the annual rainfall and all of the extreme events can be attributed to these atmospheric rivers. These cause a number of deaths and a lot of property damage. These storms are caused by low-level jets along the front of warm sectors of winter cyclones in the eastern North Pacific (Dettinger et al., 2011). They carry very large amounts of water vapor in a long narrow path. They are often more than 2000 km long, but only a few hundred km wide, and they are concentrated in the lowest 2.5 km of the atmosphere. On satellite maps, they stretch from California back to Hawaii.

Rivers

San Diego County has several rivers (Figure 3.1). They are (from north to south) San Luis Rey River, Escondido Creek, San Dieguito River, Los Penasquitos Creek, San Diego River, Cholas Creek, Sweetwater River, and the Otay River.

FIGURE 3.1 San Diego watersheds. Several rivers drain the region and flow into San Diego Bay or directly into the Pacific Ocean (Map courtesy of US Geologic Survey).

They vary in length and volume. Most run east to west and are more or less linear. Most only flow in the rainy season, but a few have water year-round. The course of the San Diego River was changed so that it now flows only into Mission Bay rather than San Diego Bay. Only the Sweetwater and Otay Rivers and Chollas Creek flow into San Diego Bay. All of these rivers are much less controlled than those in Los Angeles, but all of the rivers have dams at some point. The Tijuana River is an important component of the area, but it flows along the U.S./Mexico border and then directly into the Pacific.

Since the arrival of humans, the run-off from streams and rivers into the Bay or the Ocean has changed. Now the areas near the outflow can be polluted with sediments, toxic chemicals, pathogens, and fertilizers and sewage. The fertilizer and sewage can result in blooms of algae, which themselves sometimes contain toxins.

Aquifer

The San Diego Formation Basin contains a large aquifer that lies about 30 meters under Imperial Beach, Chula Vista, National City, and southern portions of the city of San Diego. It is estimated to contain 960,000 acre-feet of water. Unfortunately, the water is brackish. The brackish water is thought to result in the time when sea levels were much higher than current levels. Of course, the aquifer is very close to the Pacific, and so, that is always a potential source of additional saltwater. Chloride concentrations in the aquifer are 57–39,400 mg/L (Anders et al., 2014). Analysis of the isotopes in the water indicate that the relatively shallow regions of the aquifer were recharged thousands of years ago and that seawater entered in premodern times.

TOO LITTLE WATER

California and the Bay region have experienced periods of wet and dry throughout its existence. The years 1861–62 featured floods in San Diego. The years just after that are referred to as the Great Drought. During the fall and winter of 1862–63, only 9.8 cm of rain fell in San Diego County. The next year had only about 13 cm. Those years of severe drought virtually destroyed the cattle industry in California. In recent years (2011–14), the state has suffered through a multiyear drought with the lowest annual precipitation on record and the lowest in the last 1200 years (Griffin and Anchukaitis, 2014). These have resulted from a change in the usual storm track to the north that diverted rain to other areas.

Global warming has been a significant factor in the drought. Higher temperatures increase the rates of evaporation and intensify droughts. California depends on heavy snows in the Sierras in the winter that melt and runoff during the summer.

In the past, the people living in the Bay regionhad little ability to deal with an extended drought. Warmer temperatures and droughts occurred from A.D. 800 to 1350 as determined by analyses of pollen, tree rings, carbon isotopes, and fire scars on trees (Pilloud, 2006). The climate changes also stressed the Native American populations and may have caused social dislocations. Using two methods, Griffin and Anchukaitis (2014) looked at the history of drought in California. First, they examined tree ring chronologies, which is a well-established method for determining dry and wet years. Tree rings are a good proxy for determining temperatures and precipitation. Thin rings indicate less water, and thick rings plenty of water. For example, scientists using this method found that, in 200 AD, a drought in the US Southwest lasted for 50 years. Second, they used their own measurements of recent tree rings in blue oaks (*Quercus douglasii*) at four sites. The oaks are exceptionally good models for determining strong moisture signals. The researchers found that the lack of moisture is not unprecedented in California history, but they also suggest that the severity of the drought is intensified by the higher temperatures that are becoming more common with global warming. Stine (1994) used another method. He examined tree stumps that were exposed when the water level of Mono Lake decreased in the 1990s. Using radioactive carbon dating, he charted differences in water levels over time. With this method, he found two very long droughts. One began in the 9th Century and lasted 200 years. A second began in the 13th Century and lasted 150 years.

California already has extreme demands on the limited water resources available. A very real concern is that the water and agricultural infrastructure in the Bay Area and California have been built in what was an unusually wet period. If the West is normally much drier than it has been in recent years, then the area might be in for a difficult transition to deal with much less water. Brunelle and Anderson (2003) examined fossil pollen and charcoal from sediments from Siesta Lake in California to build a record of climate and fire during the Holocene. Their results are consistent with other methods (tree rings and hydrology) of determining climate for the last 1000 years. They found that the incidence of fires was much greater during the early Holocene when the sun was much brighter than it was during droughts in the "Mediaeval Warm Period," which had the longest droughts in the last 1000 years. They conclude that global warming will result in an increase in droughts and in the associated wildfires. Diffenbaugh et al. (2015) found that there have been more drought years in the past two decades than in the previous century and that the probability of conditions associated with droughts is increasing. Additional global warming over the next years is likely to cause even more events.

EFFECTS OF WATER ON LAND

Landslides are an agent of major change in mountainous areas, even greater than erosion, and either shallow or deep-seated, they are a particular hazard. Wildfires that clear off the vegetation leave hills more vulnerable without roots to add stability to the soil on slopes. That risk is enhanced if the ground has been saturated by previous rain.

Landslides are also very costly. In the period 1925–75, each year, landslides cost three times more than the combined cost of earthquakes, floods, tornadoes, and hurricanes (Jahns, 1978). They can occur anywhere, but they are most active in more mountainous areas. Shearer et al. (1983) examined the incidence of landslides in San Diego County, except for the coastal bluffs and Camp Pendleton. For the two rainy years 1978–80, the total damages in the county amounted to nearly $20 million.

Severe winter storms result in more landslides. Recently, the state adopted ARkStorm, a statewide emergency planning scenario for extreme storms. Wills et al. (2016) combined information from ARkStorm with data on areas of greatest susceptibility to landslides to produce a model that can better predict where landslides might occur.

Drought tends to reduce the number of landslides. Bennett et al. (2016) found that earth movements at less than 15 meters did not slow down in the drought during the great drought in Northern California in 2012–15. They thought that individual storms have a greater effect than expected on landslides or groundwater conditions and vegetation could influence events. However, the number of earth movements of more than 15 meters was definitely decreased. Landslides occur at different rates. The most dangerous move rapidly (tens of meters per second), and slower, less dangerous ones move as slowly as millimeters per year (Handwerger et al., 2019). They also showed that even the slow-moving landslides in Northern California greatly speeded up in the record rain year of 2017.

Wave action along the coast creates seacliff erosion. In California, about 86% of the coast is actively eroding. Many communities have built on those cliffs, and the waves can erode those cliffs and put the housing and other infrastructure at risk. Understanding how that erosion works is important for planning purposes. The cliffs in San Diego County are steep and 5–115 meters high. Benumof et al. (2000) examined erosion on sea cliffs caused by wave action. By measuring the force of the waves at several heights above the water, they found that the wave energy was actually secondary to the material composition of the cliff in determining the rate of erosion.

Nevertheless, erosion by wave action has been a significant factor in defining the California coast. For example, the La Jolla Cove caves are a well-known geologic feature of San Diego. The caves are in the face of a sandstone cliff about 60 meters high. The sandstone dates from about 75 million years ago in the Cretaceous period. They were carved by the pounding of waves on the cliffs.

CONCLUSION

Water is not as powerful as a force as plate tectonics. Nevertheless, water has had an enormous effect of the building of the San Diego Bay region. It did not build the basic structure of the Bay, but it refined the shores and mountains and provided material for the land.

REFERENCES

Anders R, Mendez GO, Futa K, Danskin WR (2014) A geochemical approach to determine sources and movement of saline groundwater in a coastal aquifer. *Groundwater* 52: 756–768.

Atwater BF, Hedel CV, Helley EJ (1977) Late Quaternary depositional history, Holocene sea-level changes, and vertical crustal movement, Southern San Francisco Bay, California. Geological Survey Professional Paper 1014.

Barberopoulou A, Legg MR, Uslu B, Synolakis CE (2011) Reassessing the tsunami risk in major ports and harbors of California I: San Diego. *Natural Hazards* 58: 479–496.

Barth NA, Villarini G, Nayak MA, White K (2016) Mixed populations and annual flood frequency estimates in the western United States: The role of atmospheric rivers. *Water Resources Research* 53: 257–269.

Bennett GL, Roering JJ, Mackey BH, Handwerger AL, Schmidt DA, Guillod BP (2016) Historic drought puts the brakes on earthflows in Northern California. *Geophysical Research Letters* 43: 5725–5731.

Benumof BT, Storlazzi CD, Seymour RJ, Griggs GB (2000) The relationship between incident wave energy and seacliff erosion rates: San Diego County, California. *Journal of Coastal Research* 16(4): 1162–1178

Bray NA, Keyes A, Morawitz WML (1999) The California current system in the Southern California Bight and the Santa Barbara Channel. *Journal of Geophysical Research* 104: 7695–7714.

Brunelle A, Anderson RS (2003) Sedimentary charcoal as an indicator of late-Holocene drought in the Sierra Nevada, California, and its relevance to the future. *The Holocene* 13: 21–28.

Chenoweth M, Landsea C (2004) The San Diego hurricane of 2 October 1858. *Bulletin of the American Meteorological Society* 85: 1689–1697.

Cordeira JM, Stock J, Dettinger MD, Young Am, Kalansky JF, Ralph FM (2019) A 142-year climatology of Northern California landslides and atmospheric rivers. *Bulletin of the American Meteorological Society* 100: 1499–1509.

Dettinger MD, Ralph FM, Das T, Neiman PJ, Cayan DR (2011) Atmospheric rivers, floods and the water resources of California. *Water* 3: 445–478

Diffenbaugh NS, Swain DL, Touma D (2015) Anthropogenic warming has increased drought risk in California. *Proceedings of the National Academy of Sciences* 112: 3931–3936.

Dolby GA, Hechinger R, Ellington RA, Findley LT, Lorda J, Jacobs DK (2016) Sea-level driven glacial-age refugia and post-glacial mixing on subtropical coasts, a palaeohabitat and genetic study. *Proceedings of the Royal Society B: Biological Sciences* 283: 20161571.

Griffin D, Anchukaitis KJ (2014) How unusual is the 2012–2014 California drought? *Geophysical Research Letters* 41: 2014GL062433.

Handwerger AL, Fielding EJ, Huang M-H, Bennett GL, Liang C, Schulz WH (2019) Widespread initiation, reactivation, and acceleration of landslides in the northern California Coast Ranges due to extreme rainfall. *Journal Geophysical Research: Earth Surface* 124: 1782–1797.

Jacobs DK, Haney TA, Louie KD (2004) Genes, diversity and geologic process on the Pacific coast. *Annual Review of Earth and Planetary Sciences* 32: 601–652.

Jahns RH (1978) *Landslides in Geophysical Predictions*. National Academy of Sciences, Washington, DC, pp. 58–65.

Otvos EG (2014) The last interglacial stage: Definitions and marine highstand, North America and Eurasia. *Quaternary International* 383: 158–173.

Pilloud MA (2006) The impact of the medieval climatic anomaly in prehistoric California: A case study from Canyon Oaks, CA-ALA-613/H. *Journal of California and Great Basin Anthropology* 26: 179–191.

Shearer CF, Taylor FA, Fleming RW (1983) Distribution and costs of landslides in San Diego County, during the rainfall years of 1978-79 and 1979-80. U.S. Geological Survey Open-File Rep. 83–582, USGS, Reston, VA.

Stine S (1994) Extreme and persistent drought in California and Patagonia during mediaeval time. *Nature* 369: 546–549.

Wills C, Perez F, Branum D (2016) New method for estimating landslide losses from major winter storms in California and application to the ARkStorm scenario. *Natural Hazards Review* 17(4). DOI: 10.1061/(ASCE)NH.1527-6996.0000142.

4 Geomorphology of the San Diego Region

INTRODUCTION

Geomorphology is the science of how the surface topography of the earth is a result primarily from the interactions of the environment upon the underlying geology (Hunt 1988). The resulting landscapes reflect not only the chemical and physical composition of the underlying rocks, but also the effect of water and weathering of those rocks, atmosphere, and hydrosphere; the soils created from the rocks by the presence and flow of water; the microbes in the soils; the botanical diversities that are successful in the environment; and the combined influence of herbivores and predators within the ecological community, namely the biosphere. Finally, the effect of humans also may contribute significantly to the landscape and thereby may undermine some of the balancing forces that had been in effect for millennia.

This chapter describes a few examples of the geomorphology of regions within the San Diego region that represent the results of different interactions between the environment, humans, and the earth's surface and which have created the landscapes we see today. We summarize the eight processes that affect geomorphology as they relate to the San Diego region: aeolian, biological, fluvial, glacial, hillslope, igneous, tectonic, and marine. We will describe a number of localities in the San Diego region that demonstrate either a particular process or a combination of processes. These localities are of interest when viewed in the context of the San Diego region, including how the underlying rock influences the overlaying vegetation and how past volcanism and plate tectonics result in the structures of the hills and mesas east of the Rose canyon fault. We also summarize how precipitation and water flow, estimated over the past 10 million years, have affected the geomorphology of the San Diego region.

Geomorphological processes generally fall into three groups: (1) the production of rock and mineral fragments (regolith) by weathering and erosion, (2) the transport of that material, and (3) its eventual deposition and interaction with the underlying surface (Derbyshire et al., 1979). The processes are further categorized as follows.

Aeolian

Named for Aeolus, the Greek god of the wind, these are wind-generated geologic processes. Wind is characterized as the movement of air between atmospheric

DOI: 10.1201/9780429487460-4

high-pressure and low-pressure systems, in particular at the interface between the atmosphere and the surface of the Earth. A greater difference in pressure between the two systems results in a greater velocity of the wind. Depending upon the wind velocity, there will be a number of different effects that wind may have upon the underlying rocks, sand, silts, minerals, and the biosphere. Silts and sands are generally defined according to size, silts being smaller than 0.0625 mm down to 0.004 mm and sands being larger than 0.0625 mm up to 2 mm (ASTM 1985). The effects can be transport, erosion, and deposition of rocks, sand, and minerals as well as the consequences upon vegetation and precipitation.

For wind velocities less than 10 $km.h^{-1}$ (~5 $m.s^{-1}$ at 1 meter above the surface) sand particles remain in place (Sloss, Hesp, and Shepherd 2012), however at greater wind velocities (≥5 $m.s^{-1}$ at 2.4 meters above the surface) sand particles may be transported many tens or hundreds of meters; in addition, greater wind velocities can transport proportionally more sand particles than at lower velocities (Webb et al., 2016). Coarse sand particles, usually a feldspar or quartz, both high in silica, having more abrasive properties than an equivalent softer rock (for example, gypsum, apatite, or other minerals that make up much of sedimentary rocks), are particularly important in the process of erosion, whereby the force of the sand particle upon a rock or mineral surface is sufficient to break off fragments of the rock or mineral, either to be carried further by the wind or deposited in the vicinity, where the fragments may later be subjected to fluvial action (see "Fluvial" below). Interestingly, vegetation cover can also influence the local wind velocity and erosion and deposition rates (Webb et al., 2016).

Biological

Biogeomorphological processes, perhaps unexpectedly to the layperson, have probably greater influence upon geomorphology of the land and the marine/freshwater environments than any other processes. First and foremost are those processes relating to the underlying geology, the subsequent weathering of those rocks by aeolian processes to create a fragmentary mineral layer, and the microbes and plants that use those minerals to grow and propagate; their waste products add to the biological and chemical detritus that, combined with the fragmentary minerals and chemical weathering, form the soil. The soil depends upon (1) the type of minerals released from the underlying rock, (2) the environmental conditions, including heat/cold, wet/dry conditions, and oxygen availability, and (3) the resulting ecosystem that is supported by that particular soil and environment.

An example of how the biota of a location can modify landform changes are the plants and animals present in marshlands, wetlands, or estuaries, for example, eelgrass and oysters. Their presence plays an important role in slowing and reducing the force of tides and storm-waters, thereby slowing the rate and effects of erosion by marine geomorphological processes (Jennings, 1996; SDUPD 2018).

Second, zoogeomorphology, whereby animals influence the form of the land is an important process. Examples include beaver dams that modify the flow of water and sediment across the land, the action of burrowing animals, digging for tubers and roots, the formation of nutrient-rich environments left by the roots and above-ground parts of trees and shrubs both before and after death, can affect the way that the soil components are transferred from one layer to another, thereby providing new environmental conditions for organisms to take advantage of. Another example is that of oyster reefs (*Ostrea lurida*) which may reduce the force of waves and wind action upon the erosion and flooding (SDUPD, 2018).

Third, and perhaps less obvious, is that the combined ecosystem can influence the balance of atmospheric carbon dioxide, which ultimately can modulate the climate. One notable exception to the large influence of biological activity upon the geomorphology of a region is to be found in Antarctica, but of course, this does not relate to the San Diego region.

Fluvial

Fluvial refers to any moving or stationary water body on the terrestrial landscape, such as creeks, streams, and rivers. Beginning from a spring or as meltwater from a glacier or snow, the water seeks the lower altitudes on its course its destination, either a lake or the sea. As it flows, it will erode the enveloping stream-bank and its bed, forming a V-shaped valley; the amount of erosion will depend upon its velocity downhill. Greater velocities result in increased erosion, which translates into more geomorphological variation over time; fast-moving streams usually have a rather more straight path, than those slower-moving waters but will deposit the sediment when suddenly slowed; alluvial fans on the sides of mountainous cliffs are an example of such flow and deposition. As the terrain becomes less steep, the water velocity slows, thus reducing the amount of bank and bed erosion and the path of the river may meander. Greater velocities upstream also result in the amount of sediment that may be transported and thus large volumes of sediment may be deposited as the river slows to a meandering phase. This in turn may build up the surface of the land through which the river flows, and may result in changes in the ecosystems and landscape. The availability of new nutrients may encourage other species to proliferate in that environment and the slow pace and slower erosional rate may result in the formation of ox-bow lakes and fluvial terraces, a botanically diverse river system that provides additional biological niches for other animals and plants.

The underlying topography of the San Diego area, the mix of highlands, mesas, valleys, and flood plains are all impacted by the rivers and streams which originate in the hill country or the mountains to the east.

Glacial

Glaciers are not a significant process within the geomorphology of the San Diego Area. The movement of glaciers down an old river valley slowly erodes the valley sides and the floor producing rock debris and creating a U-shaped valley; when the climate warms up, the glacier melts and retreats up the valley, leaving the rock debris on the surface, termed moraine. The San Diego Area was not subjected to direct effects of glaciation during the last ice age (115 to 15.2 kya) (kya: thousand years ago) and the Younger Dryas cooling (13 to 11.7 kya); The only evidence of glaciation in southern California has been found in the San Bernardino Mountains, about 200 km to the north of San Diego, and thus likely had a little topographic effect upon the San Diego region (Sharp, Allen, and Meier, 1959).

Hillslope

Soil, eroded minerals, sand, and rock, will move down a slope as creep and accumulate at the base in the valley; the slope surface can be anything from essential vertical to almost flat and the angle will determine the rate of creep. The moment that creep begins depends upon (1) the rate of weathering of the rock; (2) the amount of water present within the soils; and (3) the composition of the underlying rock. Ongoing hillslope processes will change the topology of the hill's surface, resulting in a steadily-increasing base height, and will further retard the rate of creep towards the valley floor. In addition, animal activity as reported above, may also affect the rate and onset of creep.

Historically in California, in years where there are both short-term, intense storms and above-average annual rainfall, additional factors influence the occurrence of hillslope erosion (landslides). The first was new construction and development where the hillslopes are modified by cutting and filling. The second is that of the long-term dry period that began following the Second World War, coincident with the increase in development. The third is apparently the nature of the underlying minerals and rocks (Shearer, Taylor, and Fleming, 1983). In one study, conducted in 1974 by Lough, there were notable concentrations of landslides in coastal areas generally to the west of US interstate 805 (I-805) and north of California State Highway 274 to Del Mar, southwestern Poway, and the Fletcher Hills area (Lough, 1974). By comparison, a subsequent study of the 1978–80 period showed landslides occurring mostly in the southern part of the coastal area and the Fletcher Hills area (between La Mesa and El Cajon), but nearly absent near Poway. Of interest, of the nine landslides that caused damages in excess of $1 million, seven were in the Friars Formation and two in the Calavera Hills' Santiago Formation in the northern part of San Diego County (Shearer, Taylor, and Fleming, 1983; MCC ND). As will be seen in Chapter 5, the Friars Formation comprises sandstone and claystone, the latter resulting in a more friable bedrock. The underlying Scripps Formation and the overlying Stadium Conglomerate form a mixture of mainly sandstone (with inter-bedded cobbles-conglomerate) and a sandstone/cobblestone matrix, respectively. The Santiago Formation comprises very soft sandstone inter-bedded with siltstone and mudstone. Both formations are prone to significant slumping (CDMG, 1975).

Igneous

Igneous processes, defined as the result of volcanic activity (both eruptive and intrusive), may catastrophically alter the landscape, thereby re-setting the paths of flowing water, migration routes of birds and animals, and delay the establishment of the botanical ecosystem for many years. One hundred and fifty kilometers to the east of the San Diego region lies the Salton Buttes, a dormant volcanic area that appears to have last erupted about 2000 years ago (Schmitt et al., 2012; Wright et al., 2015). The Salton Buttes includes five volcanoes that today are found near the southern portion of the Salton Sea; the Salton Sea is a remnant of a prehistoric lake which alternately dried and refilled throughout the Holocene Epoch, driven by the differential flow patterns of the Colorado River to the east (Ewing, Kocurek, and Lake, 2006) and it marks the boundary between the Pacific and North American Plates. The area is now considered to be volcanically active and may erupt at any time (Schmitt et al., 2012; Wright et al., 2015). This could cause a massive disruption to the economy and population of the San Diego region as well as the populated areas to the east in Arizona. The 156 meter Mount Calavera, near Carlsbad, 60 km north of San Diego, is the remains of a volcanic plug, the core of a volcano that last erupted from between 22 to 8.7 mya (mya: million years ago) (Day et al., 2019). The core comprises mainly fine-grained (aphanitic) dacitic felsic igneous rock derived from plutonic granodiorite (Turbeville, ND). The surrounding volcanic cone and crater have since eroded.

Tectonic

These processes are the result of the constant motion of the continental and oceanic plates, as described in Chapter 2. The resulting constant movement, which can be instantaneous or other hundreds of years, causes the relationship between adjacent ecosystems/geomorphology systems to be in constant flux, and can be observed as differences in vegetation and the animal life they support. Tectonic activity, such as earthquakes, may cause the upper portions of the crust to rise or collapse, which leads to more changes in the environment, fluvial flow, and weathering.

The San Diego Bay region is part of the Pacific plate and so has been subjected to not only periodic deposition by marine sediments for more than 50 million years, but also is under constant seismic activity due to the northward movement of the Pacific plate. This has resulted in buckling of the underlying crust and subsequent weathering to create the characteristic canyons and mesas north and east of the San Diego Bay. In addition, the Rose Canyon Fault runs straight through downtown San Diego on a roughly north-south axis, and to the west, the Point Loma Fault, a short north-south normal fault beginning on Point Loma and extending northwards towards La Jolla. Together they form a nested graben (see Chapter 2), which has contributed to formation of the San Diego and Mission Bays, as well as the Mission Hills, La Jolla Mesa, La Jolla Heights, and Torrey Pines. It is upon this graben that much of downtown San Diego, Coronado Island, the airport and naval base, the beach communities, and La Jolla, have been built. The eastern edge of the Rose Canyon Fault is bounded in part by an escarpment, upon which Old Town San Diego was built.

Approximately 125 km to the east, just beyond the Laguna Mountains, lies the Elsinore Fault, part of the fault complex that runs from the Gulf of California through eastern San Diego and Imperial Counties, dominated by the San Andreas Fault. The last major earthquake on the Elsinore fault was on May 15, 1910, and it has been estimated that the interval between major ruptures is 250 years (SCEDC, 2021).

Marine

Marine processes are those of the action of waves, marine currents, and seepage of fluids, including seawater mixed with decomposing organic material, through the seafloor. In the San Diego Bay region, the long line of predominantly sandstone cliffs and bluffs that run intermittently from north to south along the Pacific coast from the Laguna Beach in the north to La Jolla in the south typify the ongoing onslaught upon the fragile surface rock formations by the action of waves. The San Diego Bay region is noted for the frequency of large amplitude waves (think surfing) that are generated thousands of kilometers away in storms from the eastern Pacific and which are undermining many coastal homes that were once hundreds of meters from the cliff tops. As mentioned above, the tectonic forces have resulted in uplift and folding of the underlying rock, which, followed by fluvial action, has resulted in large areas of marshland and lagoons protected behind sand bars that run along the coast. The marshlands contain brackish water whereas the lagoons are predominantly freshwater. The lagoons are an important resource for seabirds, particularly diving birds that must rinse away the seawater from their feathers after fishing.

OVERVIEW

The San Diego Bay region comprises a wide range of topographies that are distinct to the region.

The majority of the landscape's form is due to the past and ongoing tectonic activity, the nature of the underlying bedrock, as well as being influenced by climate and the weather. The region is dominated by an approximately 4.5 to 10 km-wide graben trending north-south which is bounded by the broad plain of the San Diego River to the east, and a range of hills and mesas averaging 150–200 meter in height to the north-east, with small mountain ranges farther to the east. The dry mesas are intersected in an east-west trend by seasonal rivers and streams that result in deep canyons providing shelter and sustenance to diverse ecosystems.

One of the notable natural features of the region are the unique weathered bluffs north of La Jolla Cove and continuing north to just beyond Torrey Pines State Beach and a short distance east of the Rose Canyon Fault, which at that point is submerged under the Pacific Ocean. They largely comprise the Del Mar Formation (marine mudstone and siltstone with sandstone beds up to 3 meter thick), Torrey Sandstone (marine sandstone), both dating from the middle

Eocene (about 45 to 49 and 49 mya, respectively) and, separated by an unconformity, the Lindavista Formation, red siltstone and sandstone of the Pleistocene dating from about 1 to 1.5 mya (MCC, 2021; Vierra, Flynn, and Bloeser, 2017; CDMG, 1975; Thorbjarnarson, ND) (Figure 4.1). The Torrey sandstone and Torrey sandstone are strong enough to resist normal wave action, but during winter storms, they are battered with hard beach cobbles thrown by the waves against the cliff base, gradually eroding them. When the undercut is too large, the overlying rocks collapse, and a new vertical cliff-face is created. The Lindavista sandstone is the caprock and also hard but is very thin so is more prone to surface erosion. The cliffs are also intercut by streams, some of which have often been adapted to be pathways from the mesa to the beach, Black's Beach access is a good example. Mudslides from the surface layers at the top of the cliffs also contribute to further erosion of the cliffs away from the ocean. Roots of plants and lichen growing on the cliff-face, can physically and chemically break up the rock; burrowing animals may also allow water to penetrate cracks in the rock and so erode further (TPSNR, 2021).

Another natural feature is Point Loma, and which runs southward from Sunset Cliffs to the Cabrillo State Marine Reserve, approximately six km. As noted above, Point Loma is separated from the lower plane to its east by the Point Loma Fault, which undoubtedly had a role in its geomorphology. The underlying hard rocks are the Point Loma formation (Late Cretaceous inter-bedded sandstone, shale, and overlying siltstone) overlain by the Cabrillo Formation (Late Cretaceous sandstone and conglomerates), both date around 66 to 72 mya); the Cabrillo Formation is, in turn, overlain by the Lindavista Formation, dating to between 1 and 1.5 mya. The formations run southward from La Jolla Cove to the southerly tip of Point Loma (NPS, 2020).

The San Diego River presently drains into Mission Bay, resulting in the steep-sided and relatively narrow Mission Canyon extending east towards its source in the Cuyamaca Mountains, about 70 km distant. Notable peaks within the region include Carmel Mountain (130 m); Cowles Mountain, the highest point in the city of San Diego (485 m); Black Mountain (475 m); Mount Helix (418 m); El Cajon Mountain (1,112 m); McGinty Mountain (665 m); Mount Soledad (251 m); and Mount Woodson (878 m).

The region is interspersed with many giant granite boulders. The geology of central San Diego County is largely comprised of a Cretaceous gabbroic and granitic batholith (105 to 120 mya), covering approximately 44% of the county (USGS, ND; Walawender, ND). These massive rocks originated from the granitic bedrock and have been exposed to arid-climate weathering and tectonic forces that cracked the batholith and the resulting masses, the mountains of western San Diego County, were slowly broken down and rounded by temperature fluctuations and by fluvial and aeolian processes (Walawender, ND). Woodson Mountain near Poway is a typical example of such geomorphic structures.

Coronado Island, in part, created the San Diego Bay. It formed as a sand-spit filled in to the north from sediment deposited by the Tijuana River and was sheltered from wind and waves by Point Loma (late Cretaceous sandstone; the

(a)

(b)

FIGURE 4.1 The bluffs at Torrey Pines State Reserve. Top: The bluffs at Black's Beach seen from Torrey Pines Beach. Note the run-off from streams resulting in deep, narrow valleys intercutting the formations. Bottom: Strata formations visible at Black's

Beach. The three distinctive sandstone sections are visible by composition and color: The Del Mar Formation (bottom), light or pale grey and thick layers; the Torrey Sandstone (middle to top), sandy-brown and thinner, less-even layers with pockmarks towards the top. The Lindavista Formation is not visible on the bluffs as it is the caprock (Photographs reproduced with permission from Sebastian Kaser).

Cabrillo Formation) (NPS, 2020). The whole island is in fact an isthmus and two islands connected by landfill that is now home to Coronado City and the US Naval Air Station. Coronado Island is connected to Imperial Beach via the Silver Strand isthmus, thus confining the San Diego Bay. Without continuous dredging, San Diego Bay would fill with sediment and become dry land with deposition from the Sweetwater and Otay rivers (Norris and Webb, 1990).

Along the coast, to the north of the main urban areas of San Diego, are natural largely fresh-water lagoons sourced by the east to west flow of the rivers and streams. They were formed as the rivers eroded the south-south ranges of hills and highlands that characterize the San Diego area forming a floodplain behind the narrow estuaries. Some examples include the man-made Mission Bay, fed by the San Diego River and Rose Creek, Los Peñasquitos Lagoon, located between the Torrey Pines bluffs and the City of Del Mar, and which is fed by the Peñasquitos Creek, and, farther north in Del Mar, the San Dieguito Lagoon, sourced by the San Dieguito River, and the San Elijo Lagoon, north of Del Mar and Solano Beach, sourced by the Escondido Creek. All of these lagoons and marshes play a critical role in the biosphere of the San Diego area as a source of fresh water for sea birds, particularly diving birds such as pelicans (*Pelicanus occidentalis*), cormorants (*Phalacrocorax auritus*), and black skimmers (*Rynchops niger*), to wash out the salt from their feathers.

To the south of San Diego lies the Tijuana River Estuary, an intertidal coastal wetland, and which has a similar role. In addition, the estuary and the larger lagoons are key stopover points on the Pacific Flyway (Carlisle et al., 2009) and is an essential breeding, feeding, and nesting ground for over 370 species of migratory and native birds, including six endangered species, some of which include the California least tern (*Sterna antillarum browni*), the light-footed Ridgway's rail (*Rallus obsoletus levipes*), and Belding's savannah sparrow (*Passerculus sandwichensis beldingi*) (CSP, 2021; Lesberg and Redfern, 2018).

The lagoons and estuaries also support a wide variety of other organisms, including phytoplankton, macroalgae, and polychaetes; California cordgrass (*Spartina foliosa*); benthic invertebrates (mollusks and crustaceans); and fish, often juvenile marine fish, such as leopard shark (*Triakis semifasciata*), grey smoothhound (*Mustelus californicus*), and striped mullet (*Mugil cephalus*).

Regarding how the climate, precipitation, and fluvial action have molded the landscape of the San Diego region, fossils of plants dating from the late Pliocene Epoch (~ 3 mya) suggest that the climate was cooler and moister than the earlier Miocene, as well as compared to today (Axelrod and Deméré, 1984). In like manner, it appears that the climate from 1 mya to 10 kya was more like that of

Monterey today, which is 560 km north (Axelrod and Govean, 1996), but a period during which other parts of California were in the chills of the ice ages. The region is currently characterized as semi-arid Mediterranean and semi-arid steppe (Kottek, Grieser, Beck, Rudolf, and Rubel, 2006). Fossil evidence also exists that indicates an extremely diverse fauna of large animals that fed on the vegetation supported by greater levels of precipitation (see Chapter 5).

REFERENCES

ASTM (1985) Classification of Soils for Engineering Purposes: Annual Book of ASTM Standards, D 2487-83. *American Society for Testing and Materials* 4: 395–408.

Axelrod DI, Deméré T (1984) A Pliocene flora from Chula Vista, San Diego County, California. *Transactions of the San Diego Society of Natural History* 20(15): 277–300.

Axelrod DI, Govean F (1996) An early Pleistocene closed-cone pine forest at Costa Mesa, Southern California. *International Journal of Plant Sciences* 157(3): 323–329, JSTOR, www.jstor.org/stable/2475269; accessed February 15, 2021.

Carlisle JD, Skagen SK, Kus BE, Riper CV, Paxtons KL, Kelley JF (2009) Landbird migration in the American west: Recent progress and future research directions. *Condor: Ornithological Applications* 111: 211–225; DOI:10.1525/cond.2009.080096.

CDMG (1975), Geology of the San Diego Metropolitan Area, California, Bulletin 200, California Division of Mines & Geology, Sacramento, California, pp. 23, 30, 31.

CSP (California State Parks) (2021) *Tijuana Estuary NP Point of Interest*, California Department of Parks and Recreation, Sacramento, California, www.parks.ca.gov/?page_id=669; accessed February 10, 2021.

Day J, Koppers A, Mendenhall B, Oller B (2019) The "Scripps Dike" and its implications for mid-Miocene volcanism and tectonics of the California Continental Borderland. *Society for Sedimentary Geology* 110: 43–55, SEPM Special Publication.

Derbyshire E, Gregory KJ, Hails JR (1979) *Studies in Physical Geography*. Butterworth-Heinemann, Elsevier, Oxford.

Ewing RC, Kocurek G, Lake LW (2006) Pattern analysis of dune-field parameters. *Earth Surface Processes and Landforms* 31(9): 1177–1178.

Hunt CB (1988) *Geology of the Henry Mountains, Utah, as recorded in the notebooks of GK Gilbert, 1875–76*. 167. Boulder, Colorado, Geological Society of America.

Jennings P (1996) *Healing the Marsh, Explorations Summer 1996*, Scripps Institution of Oceanography, San Diego, California, pp. 1–15; https://scripps.ucsd.edu/sites/scripps.ucsd.edu/files/communications-content/field_attachment/2014/Healing_the_Marsh_V3n1.pdf; accessed February 19, 2021.

Lesberg RS, Redfern C (2018) ReWild Mission Bay: Wetlands Restoration Feasibility Study Report, San Diego Audubon Society, California State Coastal Conservancy, and U.S. Fish and Wildlife Service (Coastal Program); Everest International Consultants, Inc., Long Beach, California.

Lough CF, compiler (1974) County of San Diego landslide map, San Diego County, Office of Environmental Management, Environmental Development Agency, scale 1:125,000, San Diego, California.

MCC (2021) Field Trip Destinations in San Diego County, MiraCosta College, Oceanside and San Elijo, California, January 8, 2021; gotbooks.miracosta.edu/fieldtrips/index.html; accessed February 9, 2021.

MCC (ND) Local Geology; field trip webpage, MiraCosta College, Oceanside and San Elijo, California; http://home.miracosta.edu/jturbeville/igneous%20fieldtrip/introigneous.htm; accessed February 17, 2021.

Norris RN, Webb RW (1990) *Geology of California* (2nd edition). John Wiley & Sons, New York, pp. 298–299.

NPS (National Park Service) (2020) Geology of Cabrillo National Monument. Nps.gov/cabr/learn/nature/geology.htm#; accessed February 6, 2021.

SCEDC (2021) (Southern California Earthquake Data Center); Elsinore Fault Zone; scedc.caltech.edu/earthquake/elsinore.html; accessed February 6, 2021.

Schmitt AK, Martin A, Stockli DF, Farley KA and Lovera OM (2012), (U-Th)/He zircon and archaeological ages for a late prehistoric eruption in the Salton Trough (California, USA). *Geology* 41(1): 7–10.

SDUPD (2018) Draft Mitigated Negative Declaration for the E Street Marsh Living Shoreline Project, UPD #MND-2018-010, Aspen Environmental Group/San Diego Unified Port District, San Diego, California; https://pantheonstorage.blob.core.windows.net/ceqa/E-Street-Marsh-Living-Shoreline-Project-Draft-MND.pdf; accessed February 19, 2021.

Sharp RP, Allen CR, and Meier MF (1959) Pleistocene Glaciers on Southern California Mountains. *American Journal of Science* 257(2): 81–94.

Shearer CF, Taylor FA, Fleming RW (1983) Distribution and Costs of Landslides in San Diego County, California, During the Rainfall Years of 1978-79 and 1979-1980, Open-File Report 83-582, United States Department of the Interior, Geological Survey, Reston, Virginia.

Sloss CR, Hesp P, Shepherd M (2012) Coastal dunes: Aeolian transport. *Nature Education Knowledge* 3(10): 21.

Thorbjarnarson, K (ND) Introduction to Encinitas Geology, San Diego State University, San Diego, California, sci.sdsu.edu/geologyof/Encinitas/encgeo2.pdf; accessed February 9, 2021.

TPSNR (Torrey Pines State Natural Reserve) (2021) Nature Center geology webpage; https://torreypine.org/nature-center/geology, accessed February 8, 2021.

Turbeville, JH (ND) The Geology of Calavera Hills, North San Diego County, California: http://home.miracosta.edu/jturbeville/calavera%20hills%20%20final.doc; accessed October 22, 2020.

USGS (ND) Geological units in San Diego County, California, USGS website, US Department of the Interior, https://mrdata.usgs.gov/geology/state/fips-unit.php?code=f06073; accessed February 11, 2021.

Vierra EJ, Flynn BA, Bloeser B (2017) Depositional Processes and Facies of the Delmar and Torrey Sandstone Formations, Solana Beach, San Diego. AWG Field Trip Guide, Association for Women Geoscientists, www.geology.sdsu.eud/wp-content/uploads/2017/09/AWGfieldTripGuide2017.pdf; accessed February 9, 2021.

Walawender MJ (ND) Geologic History of San Diego, webpage published by San Diego NaturalHistory Museum and San Diego State University, San Diego; http://faculty.sdmiramar.edu/gbochicchio/Geologic%20History%20of%20San%20Diego%20County.pdf.

Webb NP, Galloza MS, Zobek TM, Herrick JE (2016) Threshold wind velocity dynamics as a driver of Aeolian sediment mass flux. *Aeolian Research* 20: 45–58.

Wright HM, Vazquez JA, Champion DE, Calvert AT, Mangan MT, Stelten M, Cooper KM, Herzig C, Schriener A, Jr. (2015) Episodic Holocene eruption of the Salton Buttes rhyolites, California, from paleomagnetic, U-Th, and Ar/Ar dating. *Geochemistry, Geophysics, Geosystems* 16(4): 1198–1210.

5 Early Biology of the San Diego Region

This chapter describes the changes in the flora and fauna of the San Diego region since the dawn of the region we call the San Diego Metropolitan Area. The majority of the underlying rocks of the San Diego region have been deposited in relatively recent times, in terms of the age of the Earth. Apart from some probable Paleozoic metamorphic rocks, the oldest rocks are marine sediments that date from the late-Jurassic period, about 165 million years ago (mya) and mainly contain the remains of plankton, ammonites and belemnites, mollusks, and sea urchin with some terrestrial forms swept out to sea from the eastern landform. Due to tectonic uplift, the shallow seas gave way to terra firma by the beginning of the Miocene (about 10 mya) and land-dwelling mammals, reptiles, birds, and amphibians become more common. Many of these would appear unfamiliar to our eyes and it was not until the beginning of the Pliocene (5.33 mya) that the large mammals we recognize as being related to our contemporaries, namely mammoths, sabre-tooth tigers, lions, hyenas, wolves, camels, horses, and deer, for example, appear in the fossil record. We end the chapter summarizing the Great Extinction event that occurred towards the end of the last ice age (about 12,000 years ago) and the conditions that may have let this come about.

EVOLUTION OF THE SAN DIEGO REGION

In previous chapters, we described the geology and geomorphology of the San Diego region during the past 165 million years. This chapter is devoted to the flora and fauna that existed during those times up until the beginning of the Holocene Epoch (11,700 years ago) and that were partly influenced by the underlying rock makeup, seascapes, and landscapes that resulted from interactions between the underlying rock and minerals and the weathering elements. Those elements include wind, atmospheric vapor concentration, water, and snow precipitation, followed by formation of streams and rivers, lagoons, wetlands, mesas, and valleys. We have chosen to include the Linnaean binomial (Genus species) nomenclature wherever we are able, so that the reader may explore elsewhere for more details if they find a topic of interest.

MESOZOIC ERA

In the Mesozoic Era (165 to 66 mya) (mya: million years ago), California was actively forming, and much of it was covered with shallow inland seas in which lived a number of marine invertebrates and reptiles. Occasionally, dinosaur

DOI: 10.1201/9780429487460-5

fossils (hadrosaur and nodosaur) have been found in the marine sedimentary rocks that make up most of these deposits and were probably washed out to sea from long-gone rivers from the eastern foothills and mountains (Weishampel, Dodson, and Osmólska, 2004; Deméré, ND). During the Cretaceous Period (145 to 66 mya) foraminifers, coccoliths, ammonites, mollusks, and oysters, were present in the shallow seas (Kennedy and Moore, 1971).

As we observed earlier, for much of its geological history the San Diego region was submerged beneath shallow coastal seas. From its creation during the late Jurassic Period (about 165 mya) until more recently (30 mya), the prehistoric San Diego region was largely affected by the subterranean motion of the Farallon plate beneath that of the North American plate, and which resulted in periodic volcanic activity as well as changes in sea level caused by tectonic upward and downward motion of the continental plates. Much of the sedimentary rock laid down within the region date from the mid- to late-Mesozoic Era (165 to 66.5 mya) and for much of the Cenozoic Era (66.5 mya to the present day) and are embedded with mainly invertebrate fossils. These include mollusks such as belemnites (squid-like cephalopods), ammonites (nautilus-like cephalopods), bivalves (oysters, clams, and pelycopods, all having bilaterally symmetrical shells, that is, their shells are upon their sides), and gastropods (including freshwater snails); crustaceans, such as crabs (decapods; Bishop, 1988; Clites, 2020) and acorn barnacles (*Sessilia* spp.); brachiopods (lamp shells, filter-feeders having dorsal/ventral shells that are distantly related to mollusks and annelid worms; Cohen and Weydmann, 2005); corals; sea urchins (such as sand dollars, order Clypeasteroida); and plankton, usually the larval forms of many invertebrate and some vertebrate species, and planktonic radiolaria (protozoa having mineralized skeletons) (Clites, 2020). What is not so readily apparent to the casual observer is the presence of microfossils, again usually larval forms of marine organisms; although invisible to the naked eye, they are of extreme importance to the energy industry geologist, for the presence (or absence) of a particular genus can indicate the co-localization of fossil fuels, such as oil and gas.

These marine organisms fluctuated both in numbers as well as across species, with the belemnites and ammonites becoming extinct at the end of the Cretaceous Period (66.5 mya), and increase in the numbers of crustacean species (Armstrong et al., 2009; Nyborg, Vega, and Filkorn, 2003).

CENOZOIC ERA

While the sedimentary deposits from the Mesozoic Era are predominantly derived from marine or otherwise fluvial (rivers and lakes) conditions, the sedimentary rocks from the Cenozoic Era (from the mass extinction at 66.5 mya to the present day) comprise both aquatic and terrestrial species.

EOCENE EPOCH: 56 TO 33.9 MILLION YEARS AGO

The earliest sedimentary rocks found in the San Diego region from the Cenozoic are from the Eocene of the San Diego region which was characterized by a warm wet climate resulting in lush tropical forests (Frederiksen, 1991) and warm shallow seas where we find marine and riverine fossil deposits (Givens and Kennedy, 1976), as well as those from terrestrial vertebrates (Walsh, 1996). Towards the end of the Eocene, the climate began to cool and the north and south poles began to accumulate ice (Peterson and Abbott, 1979; Frederiksen, 1991).

The Ardath Formation and the Scripps Formation (40 to 46 mya) predominantly include interbedded shale/sandstone and an overbedded/underbedded sandstone/cobblestone conglomerate, respectively, and indicate that the predominant fauna at the time were marine mollusks, calcareous nanoplankton flora, and foraminifera. In addition, many mammalian fossils have also been identified in these two formations. Their bodies probably had been washed out to sea from the estuaries from the eastern highlands.

Examples of the larger terrestrial vertebrates include brontotheres (*Metarhinus pater*, 46.2 to 40.4 mya) and the camel-like *Merycobunodon* (46 to 44.1 mya); examples of smaller vertebrates are a dormaaliid insectivore (*Crypholestes*), *Omomys* and *Washakius* (tarsiiform insectivorous primates, some of the earliest known crown primates; 46 to 44.1 mya), *Sespedectes* and *Proterixoides* (elephant-shrews), *Dyseolemur*, another insectivorous primate, *Rapamys* (a large rodent, somewhat similar in size to a guinea pig), *Griphomys* (an insectivorous rodent), and *Simimys*, a rodent related to mice, rats, and voles (all 44.1 to 42 mya) (Stock, 1937; Walsh, 1996).

Miocene Epoch: 23 to 5.3 Million Years Ago

At the entry of the Miocene, the western margins of the Pacific plate were accumulating marine sediments, and we find a number of marine and coastal sedimentary formations within the San Diego region. The area that now comprises the San Diego metropolitan area was mostly covered by warm seas, in which were found barnacles and other shellfish, shallow-water inhabitants. By the end of this time (about 5 mya), the Farallon plate was almost entirely subducted beneath the North American plate and the Pacific plate had already accumulated marine sediments of its own and had become uplifted.

At about 10 mya, the climate changed significantly. A number of plants from the age of the dinosaurs were the same as those existing today. For example, horsetails, ferns, and redwoods.

In the late Miocene San Mateo Formation and the Plio-Pleistocene San Diego Formation (about 5.5 and 1.5 mya, respectively) are fossils of a stem-balaenopteroid, a whale, *Norrisanima miocaena* (Boessenecker, 2013). It is thought to have resembled the present-day humpback whale since it had once been included in the same genus (Deméré, Berta, and McGowen, 2005).

Also resident in the Southern California Late Miocene seas (5 mya) was a now-extinct sea cow, *Hydrodamalis cuestae*, a relative of the ancestors of the present-day Florida manatee. During the Miocene, Central America had not yet been created and the Central American Seaway would have provided a way for

Caribbean and Atlantic sea cows to migrate to the eastern Pacific. Their remains are also found in Baja California, Mexico.

Marine mammals from this epoch are rare; one example is that of *Thalassoleon* (Otariidae; eared seal) from the late Miocene-early Pliocene boundary (about 5.6 mya). It was found in rocks in the Capistrano Formation near what are now San Clemente and San Juan Capistrano, about 100 km north of San Diego, but one would expect it to have been representative of the marine fauna at that time (Deméré and Berta, 2005).

As we mentioned earlier, animals that inhabited the Miocene environments in the San Diego region are predominantly marine and riverine/estuary invertebrates, and these environments disappear from the fossil record as we enter the Pliocene Epoch (5.33 to 2.58 mya). This change in environment correlates with the uplift of the San Diego region resulting from the relatively northward-moving Pacific plate slipping along the San Andreas Fault up the western edge of the North American plate.

Terrestrial sediments were thus forming on the eastern margins of the Pacific plate and the western margins of the North American plate and we find a plethora of plants and animal fossils from Pliocene sediments.

Pliocene Epoch: 5.3 to 2.6 Million Years Ago

Typical early Pliocene animal fossils include the traditional megafauna of North America. The term "megafauna" is considered to include megaherbivores (> 1,000 kg) and megacarnivores (> 100 kg (Malhi et al., 2016). In the early Pleistocene, these megaherbivores included camels (*Camelops*), ground sloths (*Megalonyx*), glyptodonts (*Glyptotherium*), toxodonts (*Toxodon*), oreodonts (Fam. Merycoidodontidea), and mastodons (*Mammut pacificus*), all of which evolved in the Americas. Large herbivores (45–999 kg) included primitive horses (such as the three-toed *Hipparion forcei* and *Merychippus californicus*), oreodonts, a medium-sized even-toed ungulate possibly related to camels, all now extinct (Spaulding, O'Leary, and Gatesy, 2009); large carnivores (21.5–99 kg) included saber-tooth big cats (such as *Smilodon*), American lions (*Panthera atrox*), canids (*Canis lepophagus* and *C. edwardii*, possibly the ancestors of the coyote and wolf, respectively), and hyenas (*Chasmaporthetes ossifragus*). Smaller mammals included foxes (*Vulpini*), giant beavers (*Castoroides nebrascensis*), primitive ground squirrels, mustelids (for example, badgers, martens, and otters), peccaries (*Platygonus* and *Mylohyus*), raccoon-like animals (*Procyon* spp.), and birds and lizards of various species. The middle Pliocene fauna comprised beardogs, hyenas, camels, flamingos (*Phoenicopterus copei* and *P. minutus*), ground sloths, mastodons, pronghorn antelope (*Antilocapra americana*), rhinoceroses (*Teleoceras*), cougar-like cats (*Puma pumoides*), and small rodents. The late Pliocene saw the appearance of many of North America's modern animal assemblage including bison (*Bison bison*), horses (*Equus* spp.), elk (*Cervus* spp.), and moose (*Alces americanus*) (Murray, 1974).

Most of the environmental niches of today were filled, but with different animals. The terrestrial environment was dominated by mammals and birds. The largest were the elephant and mastodon relatives called *Gomphotherium*, and they lived alongside dozens of other large mammals, such as the three-toed horse *Hipparion,* the tiny pronghorn *Merycodus*, the long-necked camel *Aepycamelus,* and peccaries. There were also smaller mammals, such as rabbits, beavers, ground squirrels, foxes. Birds and reptiles shared the space. The large number of herbivores supported a significant number of predators. The dog *Borophagus* probably killed and also ate carrion. *Nimravides* was a predatory cat. The largest predator was the sabertooth cat *Barbourofelis*.

Great American Biotic Interchange

The Great American Biotic Interchange (GABI) was an important paleozoogeographic event in which animals and plants were exchanged between the north and south American continents around the end of the Pliocene Epoch. Three million years ago, tectonic forces and subduction of the Cocos plate under the Caribbean and North American plates formed the Panama isthmus between the Americas (O'Dea et al., 2016).

The GABI thus represented a north-south movement of Eutheria from North America to South America (cats, camels [e.g., vicuñas, the parent species of llamas], tapirs, and peccaries) and northward flow from South America of opossums and armadillos (O'Dea et al., 2016).

The Pleistocene (beginning about 2.6 mya) ushered in a long period of climatic instability, including successive cycles of glacial (ice ages) and interglacial (warmer periods) periods (Cohen et al., 2013). The reasons for this climatic instability are complex, but have been associated with wobble and periodic reversals of the Earth's magnetic field; fluctuations in the amount of energy from the sun, most likely due to eccentricities in the Earth's orbit around the sun; and tectonic/volcanic activity (Buis, 2020; Nordt, Atchley, and Dworkin, 2003; Foulger, 2010).

The presence of marsupials in the Americas may be seen to some as odd since marsupials are usually associated with Australian fauna. This is where an understanding of palaeontology and geology bring to light the most likely explanation. During the early Cretaceous period, around 126 mya, the landmasses of North America, Europe, and Asia (composing the supercontinent of Laurasia) were in the process of breaking up; South America and Australia were still connected by Antarctica as part of the larger supercontinent Gondwana. At that time dinosaurs were the dominant large animal group and the ancestors of the marsupials, termed metatherians, were generally small and most likely nocturnal. Unfortunately, there is scant fossil evidence for these marsupial ancestors living on all three southern continents at that time; only monotremes and their ancestors have been found (Benson et al., 2013). It is most likely that the metatherians evolved in Asia, subsequently moved into North America by way of Europe, and then, at the end of the Cretaceous (66 mya), crossed by a landbridge to South America and then onwards expanding into Gondwana (Flynn and Wyss, 1998;

Flynn, Wyss, and Charrier, 2007). Plate tectonics then caused the landbridge to migrate north-east, where it became part of the Caribbean Archipelago, thereby cutting off any further connections between the continents of Laurasia and Gondwana (Kemp, 2005; Boschman et al., 2014). There is, however, also evidence of non-placental therian fossils already in Gondwana during the mid-late Cretaceous (83.6 to 66 mya; Newham et al., 2014).

Pleistocene and Holocene Epochs: 2.6 Million Years Ago to the Present

The Pleistocene Epoch (from 2.6 million to about 11,000 years ago) includes the climatic changes that brought about the periodic ice ages, as we discussed in Chapter 2 and above. It is here that the mammoths finally enter the North American ecosystem, having evolved around 5 mya in Africa from an ancestor that also gave rise to the Asian elephant (Lister and Bahn, 2007; Krause et al., 2006). It is considered that the Columbian mammoth (*Mammuthus columbi*) in North America is descended from hybrids of either the steppe mammoth (*Mammuthus trogontherii*) or the woolly mammoth (*Mammuthus primigenius*) with the Krestovka mammoth (*Mammuthus* unk. sp.) in Siberia and the resulting hybrid(s) then entered North America over 1 mya before finally evolving into the Columbian mammoth (Lister and Sher, 2015; van der Valk, Pečnerová, and Díez-del-Molino, 2021). The woolly mammoth (*M. primigenius*) also had evolved in Siberia and subsequently migrated across Beringia during an interglacial period possibly by around 400 kya and then also interbred with the Columbian mammoth (Lister and Sher, 2015; van der Valk et al., 2021). The Pleistocene Epoch also saw the appearance of the American lion (*Panthera atrox*), American cheetah (*Miracinonyx* spp.), giant jaguar (*Panthera onca augusta*), puma (*Puma concolor*), lynx (*Lynx rufa*), dire wolf (*Canis dirus*), timber or grey wolf (*Canis lupus*), grey fox (*Urocyon cinereoargenteus*), short-faced bear (*Arctodus simus*), modern horse (*Equus ferus/caballus*), steppe bison/bison (*Bison priscus/Bison bison*), reindeer/caribou (*Rangifer tarandus*), shrub-ox (*Euceratherium collinum*), musk ox (*Ovibos moschatus*), tapir, pronghorn, elk, mule deer, bighorn sheep, and peccaries (Lorenzen et al., 2011; Oberbauer, 2018; Chimento and Dondas, 2018). Some scientists have compared it to the modern-day Serengeti in Africa (Parkman, 2006).

About this time, the area from Tijuana, Mexico, to as far as Pacific Beach and Mount Soledad in San Diego, was a large crescent-shaped bay, about the size of present-day Monterey Bay in northern California. The marine and non-marine sandstone sediments of the bay compose the San Diego Formation, which covers much of the San Diego metropolitan area (CDMG, 1975). The invertebrate fossils embedded in the sediments are a mixture of mollusks, echinoderms, and crustaceans; interestingly, a few species of clams, such as *Dosinia ponderosa* and *Mitha xantusi*, are now found to the south in the more tropical Gulf of California, suggesting that the climate then was warmer and wetter than today. Also present are the fossil bones and teeth of fish and marine mammals; the marine mammals include several species of baleen whales, toothed whales, fur seals, walruses, and sea cows, many of which are now extinct (Rugh, 1998).

During the late Pleistocene and the early Holocene, plants and animals evolved very little, but there were big changes in their populations and especially among the megafauna. The mammals included mammoths, mastodons, camels, horses, llamas, elk, tapirs, moose, and bison, along with large predators, such as the short-faced bear, saber-tooth cat, wolves, and American lions.

All of these species and more died out and were not replaced by others as was the normal expectation. The reason is not clear. Some scientists suggest that climate change was responsible (Wroe and Field, 2006; Guthrie, 2003). However, others focus on the expansion of humans and their arrival in North America. Barnosky et al. (2016) weighed these two theories. First, they found that the average temperature changes and anomalous precipitation and the velocity of those changes 132,000 to 1,000 years ago and found no correlation with the loss of mammals. From these results, they discount the effects of climate change on the disappearance of the megafauna in the early Holocene. Second, they evaluated the effects of humans on the animals. Their results suggest that early hominoids in Africa and those that left Africa (e.g., Neanderthals and Denisovans) co-evolved with the large mammals there. Once modern humans (*Homo sapiens*) began to leave Africa and spread into Australia, Northern Asia and Europe, and North and South America, they were the first humans to have contact with the mammals in those areas. Modern humans brought new tools and hunting techniques that easily and quickly overwhelmed the large mammals.

About 100,000 years ago, the Late Pleistocene Epoch began. At that time, large herds of mammoth, mastodons, camels, horses, bison, llamas, elk, and tapirs were present (Hoppe, 2004; Grayson and Meltzer, 2015, Oberbauer, 2018). Over them, all soared the condor (*Terratornis merriam*) with its wingspan of over 3 meters. The herds likely moved from inland in the winter to the shore in the summer to escape the heat and eat the more vegetation kept lush by the fog. These would have been fairly short migrations of less than 60 km.

Among the large mammals were the Columbian mammoth (*Mammuthus columbi*), which weighed more than 5 tonnes and stood 4 meters at the shoulders. Long-horned bison (*Bison latifrons*) and ancient bison (*Bison antiquus*) both grazed on grasses. Both were larger than modern bison. The long-horned bison lived alone or in small groups, but the ancient bison were herd animals. The American mastodon (*Mammut americanum*) was distantly related to elephants; it was shorter and stockier than the mammoth. The western horse (*Equus occidentalis*) was small and stood only about 1.5 meters at the shoulders. The giant horse (*Equus pacificus*) was larger. Interestingly, the large-headed llama (*Hemiauchenia macrocephala*) was a grazer and far from the Andes that are normally associated with llamas. Jefferson's ground sloth (*Megalonyx jeffersoni californicus*) was as large as an ox. Camel (*Camelops hesternus*) were herd animals of about the same size as modern camels.

Well-fed predators were there too, including the short-faced bear, saber-tooth cat, dire wolf, and American lion. The dire wolf (*Canis dirus*) was about the size of a modern-day timber wolf and probably hunted in packs as do modern wolves. The large number of dire wolves found in the La Brea Tar Pits suggests that they ate carrion as well as hunted. Recent studies of dire wolf DNA from sub-fossil remains have found that when compared with modern wolves, coyotes, and jackals, they are genetically quite distinct from that group, having split from the Eurasian canids about 5.7 mya (Perri, Mitchell, Mouton, et al., 2021). The authors suggested that this supports the taxonomic classification under the prior name *Aenocyon dirus*. The coyote (*Canis orcutti*) is now extinct. A large number of them were also found in the La BreaTar Pits, and they were probably significant predators. They were larger than their modern counterparts, but could not adjust after most of the larger mammals were eliminated. The giant short-faced bear (*Arctodus simus*) was the largest carnivore that ever lived in the New World. An adult weighed about 900 kg and was a third larger than a grizzly bear. In addition, its longer legs suggest that it was faster even than modern bears. The cat family was well represented. The scimitar cat (*Homotherium serum*) was closely related to the saber-tooth cat (*Smilodon californicus*) and hunted large animals. In addition to many of the modern cat forms, the saber-tooth cats were major predators. The American lion (*Panthera atrox*) was larger than modern-day cats in Eurasia and may have hunted in prides as with modern lions. The jaguar (*Panthera onca*), American cheetah (*Acinonyx trumani*), cougar (*Felis daggetti*), and lynx (*Lynx rufa fischeri*) rounded out this group.

The Quaternary Glaciation began about 2.58 mya and is ongoing (Marshall, 2010). From then to the present, the Earth has experienced several glacial periods. In that time, there have been several cycles of expansion and retreat of ice. In the colder parts of the cycle, ice sheets and glaciers covered a significant portion of the Northern Hemisphere. Temperatures and sea levels fell. Between those, temperatures and sea levels rose. For much of the Quaternary, the ice cycle repeated about every 41,000 years, but about a million years ago, the cycle changed to every 100,000 years. The reasons are not quite clear. In fact, the cause of the cycles themselves are not known but seem to be related to the tilt of the Earth.

By the beginning of the ice age that began about 2 mya, many of the earlier mammals, including 80 genera, had disappeared. As is almost always the case, when one genus disappears, another takes its place. Those early mammals were replaced by mammoths, dire wolves, and sabertooth cats. Some sort of extinction event seems to have eliminated many of the living organisms, but not to the extent that it could be considered a mass extinction event. The cause is unknown. Some speculate that climate change was involved. At about that time, the Earth cooled, the ice sheets at the polar caps expanded, glaciers expanded southward, rainfall was reduced, and the oceans cooled and sea levels fell. Other complications might have added to the stress on the mammals. Climate change also affected the vegetation. Grasslands gave way to a different environment more like our current one that features chaparral, multiple grasses, and many new species. Perhaps some combination of all of these was responsible for the loss of the genera. Some have speculated that the arrival of Native Americans might have contributed to the demise of many of these species. However, that hypothesis remains unproved at this point.

The last of these ice ages began 33,000 years ago and started to recede 19,000 years ago. During that time, sea levels were much lower than before, and a land bridge was revealed between Siberia and Alaska. Native Americans crossed that bridge to become the first humans in the America. They made it to the San Diego region about 15,000 years ago.

The Great Megafauna Extinction

As the last ice age began to come to an end about 15,000 years ago, almost all the megafauna of North America became extinct. The causes of these extinctions have been debated amongst scientists for many decades and have revolved from (1) adaptations (or lack of) to climate change, (2) a reduction in habitat resulting in reduced food supply for these huge mammals, (3) the appearance of man, who may have hunted them to extinction, or (4) a combination of all these factors (Barnosky et al., 2004; Martin, 1984; Alroy, 2001; Hoppe, 2004; Stuart et al., 2004; Koch and Barnosky, 2006; Sandom et al., 2014; Surovell et al., 2016). A study that modeled population dynamics of six megafauna species over the past 50,000 years suggested that climate change and population fragmentation, and for certain species, human engagement, played the major role in their extinction in North America (Lorenzen et al., 2011). The remaining medium-to-large herbivores, such as bison, elk, moose, reindeer, and pronghorn antelope, survived this period of change, for reasons as yet unknown; this may be a combination of surviving in large herds, adapting to the relatively warmer climates, and the multiplicity of food available in the woods and grasslands that came to dominate the landscape after the ice age. Musk ox survived in the Canadian and Alaskan Arctic (Lorenzen et al., 2011). In addition, populations of moose (*Alces*) and elk (*Cervus* spp.) increased towards the end of the Late Quarternary Epoch (late Pleistocene) that may have led to increased competition for both woodland and grassland resources (Guthrie, 2006). Of note, both reindeer (in Eurasia) and caribou (North America) have been hunted and/or herded by Arctic peoples for at least the past 10 to 15 thousand years (Kurtén, 1968) thus perhaps inadvertently preserving pan-regional breeding populations.

Many cases of animals going extinct or nearly extinct are well known. The Columbian mammoth (*Mammuthus columbi*) lived in the San Diego region during the Late Pleistocene. These relatives of modern elephants stood about 4 meters tall, weighed 11,000 kg with tusks almost 5 meters long. They migrated into North America 2 mya but became extinct about 11,000 years ago. Other mammals included 2-tonne bison, Western camels, giant horses, tapirs, and giant ground sloths. There were probably many American lions that lived in what is now San Diego. They weighed 340 kg and were much larger than modern African lions. The other major predator was the short-faced bear. They were 4 meters long, weighed 900 kg, and could probably run at 65 kph for just over 1 km. Other predators included dire wolves, saber-toothed cats, grizzly bears, and giant condors.

California originally had two types of bears. The black bear (*Ursus americanus*) was found as far south as Sonoma County in northern California but was never native to San Diego County. The California grizzly bear (*U. arctos californicus*) was found throughout the state. Thus, any reference to bears in early writings almost surely refers to the grizzly. The California grizzlies were larger than the grizzlies that lived in the Rockies (900 kg as compared to 700 kg). But by the mid-19th century, cattle had become a major industry in California, and the grizzlies were a problem. Farmers began killing them. Within 75 years of the discovery of gold, the California grizzly bear was extinct.

We will now present a brief description of the flora associated with the period from about 10 mya to the beginning of the Holocene Epoch, about 11,700 years ago.

Flora

In the late Miocene and early Pliocene epochs (about 7 to 5 mya), a change occurred in the ratio of plants using the C3 and C4 photosynthetic pathways. For most plants on Earth, the first product of photosynthesis is a three-carbon product. In this process, because CO_2 enters through the stomata, and when the stomata are open, the plant can lose significant amounts of water. During a drought, this is a disadvantage. Plants in hot, dry areas have evolved C4 photosynthesis, which results in a four-carbon product. This process allows plants to carry on photosynthesis with the stomata closed. C4 plants include maize, sugarcane, and sorghum. The C4 process is also favored under conditions of low carbon dioxide levels. Cerling et al. (1997) examined the teeth of grazing animals during that period and determined that a significant change in the diets of those animals indicate a change in the atmospheric CO_2 levels that favored C4 photosynthesis.

California is one of a few regions in the world that features an exceptionally large number of unique plant species. That diversity stems from the late Tertiary period and is likely a function of the state's multiple environments with different altitudes and amounts of rainfall. However, Harrison, Safford, and Wakabayashi (2004) speculated that the diversity might also be a function of the presence of the mineral serpentine. Serpentine is an ultramafic rock that produces soils with fewer nutrients (nitrogen, phosphorous, potassium, and calcium) than other soils and more toxins (magnesium, nickel, and chromium). Harrison, Safford, and Wakabayashi (2004) found that an association between the presence of serpentine and increased diversity exists, particularly in the Klamath and northern Coast Range. The age of exposure of the serpentine was an important factor. Anacker et al. (2010) further explored the influence of serpentine soils on plant diversity. They found that, once a plant lineage is specialized for its environment, it is less likely to further diversify than other plant lineages.

Plants were highly diverse by the time of the dinosaurs (Millar and Woolfenden, 2016). Conifers appeared about 200 mya. The pines were the last to develop, and they were common about 145 mya. Gymnosperms, including ferns and horsetails, expanded worldwide about 150 mya. Angiosperms, traced back to 110 to 150 mya, had expanded to become the most diverse group of plants on Earth. Angiosperms were in California at 100 to 120 mya, but their diversity can be pegged to about 55 mya. However, temperatures and humidity rose from 50 to 52 mya, and the angiosperms bloomed (no pun intended). Species found in more tropical areas were common, including avocado, palm, viburnum, magnolia, jackfruit, and figs (Millar and Woolfenden, 2016). Pollen was found for pine, walnut, hickory, and sweetgum.

At about 33 mya, the tropical species disappeared and more temperature species again became dominant. Evergreen oaks, sycamore, cottonwood, willow, redbud, barberry, cherry, ironwood, manzanita, flannel bush, sumac, and grasses were found in Contra Costa County (Edwards 2004).

Regarding plant species, at the boundary between the Miocene and the Pliocene (about 5.3 mya), alder, cherry, Christmas berry, chumico, coffee berry, dogwood, elm, flannel bush, Catalina ironwood, California lilac, magnolia, mountain mahogany, manzanita, live oak, poplar, bush poppy, swamp cypress, sumac, desert sweet, sycamore, tupelo, and willow all grew around the San Diego Bay region (Murray, 1974).

REFERENCES

Alroy JA (2001) Multispecies overkill simulation of the end-Pleistocene megafaunal mass extinction. *Science* 292: 1893–1896.

Anacker BL, Whittall JB, Goldberg EE, Harrison SP (2010) Origins and consequences of serpentine endemism in the California flora. *Evolution* 65: 365– 376.

Armstrong A, Nyborg T, Bishop G, Osso A, Vega FJ (2009) Decapod crustaceans from the Paleocene of central Texas, USA. *Revista Mexicana de Ciencias Geologicas* 26(3): 745–763.

Barnosky AD, Koch PL, Feranec RS, Wing SL, Shabel AB (2004) Assessing the causes of late Pleistocene extinctions on the continents. *Science* 306: 70–75.

Barnosky AD, Lindsey EL, Willavicencio NA, Bostelmann E, Hadly EA, Wanket J, Marshall CR (2016) Variable impact of late-Quaternary megafaunal extinction in causing ecological state shifts in North and South America. *Proceedings of the National Academy of Sciences USA* 113: 856–861.

Benson RBJ, Mannion PD, Butler RJ, Upchurch P, Goswami A, Evans SE (2013) Cretaceous tetrapod fossil record sampling and faunal turnover: Implications for biogeography and the rise of modern clades. *Paleogeography, Paleoclimatology, Paleoecology* 372: 88–107.

Bishop GA (1988) Two crabs, *Xandaros sternbergi* (Rathbun 1926) n. gen., and *Icriocarcinus xestos* n. gen., n. sp., from the Late Cretaceous of San Diego County, California, USA, and Baja California Norte, Mexico. *Transactions of the San Diego Society of Natural History* 21: 245–257.

Boessenecker RW (2013) A new marine vertebrate assemblage from the Late Neogene Purisima Formation in Central California, part II: Pinnipeds and Cetaceans. *Geodiversitas* 35(4): 815–939.

Boschman LM, van Hinsbergen DJJ, Torsvik TH, Spakman W, Pindell JL (2014) Kinematic reconstruction of the Caribbean region since the Early Jurassic. *Earth-Science Reviews* 138: 102–136.

Buis A (2020) "Milankovitsch (orbital) cycles and their role in Earth's climate." NASA https://climate.nasa.gov/news/2948/milankovitch-orbital-cycles-and-their-role-in-earths-climate/; accessed September 28, 2020.

CDMG (1975) *Geology of the San Diego Metropolitan Area, California, Bulletin 200.* California Division of Mines & Geology, Sacramento, California.

Cerling TE, Harris JM, MacFadden BJ, Leakey MG, Quade J, Eisenmann V, Ehleringer JR (1997) Global vegetation change through the Miocene/Pliocene boundary. *Nature* 389: 153–158.

Chimento N, Dondas A (2018) First Record of *Puma concolor* (Mammalia, Felidae) in the Early-Middle Pleistocene of South America. *Journal of Mammalian Evolution* 25(Suppl. 1): 1–9, DOI: 10.1007/s10914-017-9385-x

Clites E (2020) Fossils in our Parklands. University of California Museum of Paleontology, https://ucmp.berkeley.edu/science/parks/golden_gate.php; accessed 2020-09-12.

Cohen BL, Weydmann A (2005) Molecular evidence that phoronids are a subtaxon of brachiopods (Brachiopoda: Phoronata) and that genetic divergence of metazoan phyla began long before the early Cambrian. *Organisms Diversity & Evolution* 5: 253–273.

Cohen KM, Finney SC, Gibbard PL, Fan J-X (2013) International Chronostratigraphic Chart 2013, stratigraphy.org, ICS; accessed January 7, 2019.

Deméré T, Berta A (2005) New Skeletal Material of *Thalassoleon* (Otariidae: Pinnipedia) from the Late Miocene-Early Pliocene (Hemphillian) of California. *Bulletin of the Florida Museum of Natural History* 45(4): 379–411.

Deméré TA, Berta A, McGowen MR (2005) The taxonomic and evolutionary history of modern balaenopteroid mysticetes. *Journal of Mammalian Evolution* 12(1/2): 99–143).

Deméré T (ND) Dinosaurs of San Diego County, San Diego Natural History Museum, https://www.sdnhm.org/download_file/view/1823/195; accessed February 11, 2021.

Edwards SW (2004) Paleobotany of California. *The Four Seasons* 4: 3–75.

Flynn JJ, Wyss AR (1998) Recent advances in South American mammalian paleontology. *Trends in Ecology and Evolution* 13(11): 449–454; DOI: 10.1016/S0169-5347(98)01457-8. PMID 21238387.

Flynn JJ, Wyss AR, Charrier R (2007) South America's missing Mammals. *Scientific American* 296 (May): 68–75.

Foulger GR (2010) *Plates vs. Plumes: A Geological Controversy.* Wiley-Blackwell, Heboken, New Jersey.

Frederiksen NO (1991) Pulses of middle Eocene to earliest Oligocene climatic deterioration in Southern California and the Gulf Coast. *Palaios* 6(6): 564–571.

Givens CR, Kennedy MP (1976) Middle Eocene mollusks from Northern San Diego County, California. *Journal of Paleontology* 50(5): 954–975.

Grayson DK, Meltzer DJ (2015) Revisiting Paleoindian exploitation of extinct North American mammals. *Journal of Archaeological Science* 56: 177e193.

Guthrie RD (2003) Rapid body size decline in Alaskan Pleistocene horses before extinction. *Nature* 426: 169–171.

Guthrie RD (2006) New carbon dates link climatic change with human colonization and Pleistocene extinctions. *Nature* 441: 207–209.

Harrison S, Safford H, Wakabayashi J (2004) Does the age of exposure of serpentine explain variation in endemic plant diversity in California? *International Geology Review* 46: 235–242.

Hoppe KA (2004) Late Pleistocene mammoth herd structure, migration patterns, and Clovis hunting strategies inferred from isotopic analyses of multiple death assemblages. *Paleobiology* 30: 129–145.

Kemp TS (2005) *The Origin and Evolution of Mammals*. Oxford University Press, Oxford, p. 217.

Kennedy MP, Moore GW (1971) Stratigraphic relations of Upper Cretaceous and Eocene formations, San Diego coastal area, California. *American Association of Petroleum Geologists Bulletin* 55(5): 709–722.

Koch PL, Barnosky AD (2006) Late Quaternary extinctions: State of the debate. *Annual Review of Ecology, Evolution, and Systematics* 37: 215–250.

Krause J, Dear PH, Pollack JL, Slatkin M, Spriggs H, Barnes I, Lister AM, Ebersberger I, Pääbo S, Hofreiter M (2006) Multiplex amplification of the mammoth mitochondrial genome and the evolution of Elephantidae. *Nature* 439: 724–727.

Kurtén B (1968) *Pleistocene Mammals of Europe*. Transaction Publishers, Piscataway, New Jersey, pp. 170–177.

Lister A, Bahn P (2007) *Mammoths: Giants of the Ice Age*. Frances Lincoln LTD, London, UK, p. 23.

Lister AM, Sher AV (2015) Evolution and dispersal of mammoths across the Northern Hemisphere. *Science* 350: 805–809.

Lorenzen ED, Nogués-Bravo D, Orlando L, Weinstock J, Binladen J, Marske KA, Ugan A, Borregaard MK, Gilbert MTP, Nielsen R, Ho SYW, Goebel T, Graf KE, Byers D, Stenderup JT, Rasmussen M, Campos PF, Leonard JA, Koepfli K-P, Froese D, Zazula G, Stafford Jr TW, Aaris-Sørensen K, Batra P, Haywood AM, Singarayer JS, Valdes PJ, Boeskorov G, Burns JA, Davydov SP, Haile J, Jenkins DL, Kosintsev P, Kuznetsova T, Lai X, Martin LD, McConald HG, Mol D, Meldgaard M, Munch K, Stephan E, Sablin M, Sommer RS, Sipko T, Scott E, Suchard MA, Tikhonov A, Willerslev R, Wayne RI, Cooper A, Hofreiter M, Sher A, Shapiro B, Rahbek C, Willerslev E (2011) Species-specific responses of Late Quaternary megafauna to climate and humans. *Nature* 479: 359–364.

Malhi Y, Doughty CE, Galetti M, Smith FA, Svenning J-C, Terborgh JW (2016) Megafauna and ecosystem function from the Pleistocene to the Anthropocene. *Proceedings of the National Academy of Sciences* 113: 838–846.

Marshall M (2010) The history of ice on Earth. *The New Scientist*: https://www.newscientist.com/article/dn18949-the-history-of-ice-on-earth/.

Martin PS (1984) Prehistoric overkill: The global model. In: *Quaternary Extinctions: A Prehistoric Revolution* (eds. Martin PS, Klein RG) University of Arizona Press, Tucson, Arizona, pp. 364–403.

Millar CI, Woolfenden WB (2016) Ecosystems past: Vegetation prehistory. In: *Ecosystems of California* (eds. Mooney H, Zavaleta E) University of California Press, Berkeley, CA, pp. 131–154.

Murray M (1974) *Hunting for Fossils: A Guide to Finding and Collecting Fossils in All 50 States*. Collier Books, Springfield, Ohio, 97–102, 348.

Newham E, Benson R, Upchurch P, Goswami A (2014) Mesozoic mammaliaform diversity: The effect of sampling corrections on reconstructions of evolutionary dynamics. *Paleogeography, Paleoclimatology, Paleoecology* 412: 32–44.

Nordt L, Atchley S, Dworkin S (2003) Terrestrial evidence for two greenhouse events in the latest cretaceous. *GSA Today* 13(12): 4–9. DOI: 10.1130/1052-5173; accessed February 16, 2021.

Nyborg TG, Vega FJ, Filkorn HF (2003) New Late Cretaceous and Early Cenozoic decapod crustaceans from California, USA: Implications for the origination of taxa in the eastern North Pacific. *Contributions to Zoology* 72: 165–168.

Oberbauer T (2018) Prehistoric San Diego County, Part 2, California Native Plant Society and San Diego Natural History Museum, San Diego, California, Prehistoric San

Diego County, Part 2 – California Native Plant Society-San Diego Chapter (squarespace.com); accessed February 27, 2021.

O'Dea A, Lessios HA, Coates AG, Eytan RI, Restrepo-Moreno SA, Cione AL, Collins LS, de Queiroz A, Farris DW, Norris RD, Stallard RF, Woodburne MO, Aguilera O, Aubry MP, Berggren WA, Budd AF, Cozzuol MA, Coppard SE, Duque-Caro H, Finnegan S, Gasparini GM, Grossman EL, Johnson KG, Keigwin LD, Knowlton N, Leigh EG, Leonard-Pingel JS, Marko PB, Pyenson ND, Rachello-Dolmen PG, Soibelzon E, Soibelzon L, Todd JA, Vermeij GJ, Jackson JB (2016) Formation of the Isthmus of Panama, *Science Advances* 2(8):e1600883

Parkman EB (2006) The California Serengetti: Two hypotheses regarding the Pleistocene paleoecology of the San Francisco Bay Area. Retrieved from: http://www.parks.ca.gov/pages/22491/files/the_california_serengetti_pleistocene_paleoecology_of_san_francisco_bay.pdf.

Perri AR, Mitchell KJ, Mouton A, Álvarez-Carretero S, Hulme-Beaman A, et al., (2021). Dire wolves were the last of an ancient New World canid lineage. *Nature*, 591(7848), 87–91. 10.1038/s41586-020-03082-x.

Peterson GL, Abbott PL (1979) Mid-Eocene climatic change, Southwestern California and Northwestern Baja California. *Paleogeography, Paleoclimatology, Paleoecology* 26: 73–87.

Rugh S (1998) Clams of Champions: The San Diego Formation, San Diego Natural History Museum: http://archive.sdnhm.org/research/paleontology/sdform.html; accessed October 19, 2020.

Sandom C, Faurby S, Sandel B, Svenning J-C (2014) Global late Quaternary megafauna extinctions linked to humans, not climate change. *Proceedings of the Royal Society B: Biological Sciences* 281: 1–9.

Spaulding M, O'Leary MA, Gatesy J (2009) Relationships of Cetacea (Artiodactyla) among mammals: Increased taxon sampling alters interpretations of key fossils and character evolution. *PLoS One* 4(9): e7062.

Stock C (1937) An Eocene titanothere from San Diego County, California, with remarks on the age of the Poway Conglomerate. *Proceedings of the National Academy of Sciences of the USA* 23(2): 48–53.

Stuart AJ, Kosintsev PA, Higham TFG, Lister AM (2004) Pleistocene to Holocene extinction dynamics in giant deer and woolly mammoth. *Nature* 431: 684–689.

Surovell TA, Pelton SR, Anderson-Sprecher R, Myers AD (2016) Test of Martin's overkill hypothesis using radiocarbon dates on extinct megafauna. *Proceedings of the National Academy of Sciences of the USA* 113: 886–891.

van der Valk T, Pečnerová P, Díez-del-Molino D, Bergström A, Oppenheimer J, et al. (2021) Million-year-old DNA sheds light on the genomic history of mammoths. *Nature*, https://doi.org/10.1038/s41586-021-03224-9

Walsh SL (1996) Middle Eocene mammalian faunas of San Diego County, California. In: *The Terrestrial Eocene-Oligocene Transition in North America* (eds. Prothero DR, Emry RJ) Cambridge University Press, UK, Cambridge

Weishampel DB, Dodson P, Osmólska H (2004) Dinosaur distribution (Late Cretaceous, North America). In: The Dinosauria, 2nd (eds. Weishampel DB, Dodson P, Osmólska H) University of California Press, Berkeley, pp. 574–588.

Wroe S, Field J (2006) A review of the evidence for a human role in the extinction of Australian megafauna and an alternative interpretation. *Quaternary Science Reviews* 25: 2692–2703.

6 Humans Arrive

The San Diego Bay and region were created by massive tectonic, volcanic, and hydrological forces over millions of years. Humans have little ability to change the primary structure of the Bay region. However, since they arrived, humans have put their stamp on the region. Their influence has been felt in greater measure on the shorelines and rivers, but the air, water, and animals were affected to a much greater extent.

NATIVE AMERICANS

Earliest Humans and Their Lives

The first humans in the San Diego Bay region were Native Americans, and they arrived a long time ago. The earliest migrations of modern humans into the Americas are complicated and not completely understood (Waters, 2019). The traditional view holds that humans migrated across an overland route into North America, and then, they spread out around the Americas. However, this Clovis model fell apart as more evidence was found to support a model in which early humans followed the shoreline into the Americas (Figure 6.1). Many sites have been found along the shores, and many more were lost over time to sea level rise, erosion, and tectonic movements (Braje et al., 2017). Nevertheless, sites at Monte Verde in Chile show human habitation at least 14,000 years ago and maybe as much as 16–18,000.

There is evidence of very early humans in the San Diego area. The Arlington Man skeleton was found on Santa Rosa Island, one of the Channel Islands off Los Angeles, and dated to 13,000 years ago (Braje et al., 2010). Human artifacts have been found on Cedros Island off Baja California. Those remains, including the earliest fishhooks in the Americas, were dated to 13,000 years ago (Erlandson et al., 2008; Wade, 2017). Again, some of the evidence may be buried underwater now. Sea levels have risen considerably since those earliest days and covered several kilometers of land in the San Diego region that likely contained early inhabited sites.

The extent of prehistoric settlements and their effects on the environment are unclear (Scharlotta, 2015). A large number of specimens have been collected and dated, but much remains to be understood. For example, changes that seemed to happen in the early Holocene do not match the climate of that same time (Figure 1.3). Interestingly, the arrival of the bow and arrow was much later in human existence in the San Diego area, perhaps 1250 to 1600 years before present. Nevertheless, it had a great effect on the inhabitants. Hunting became much more

FIGURE 6.1 Changing California coastline. The coastline is a function of sea level. At the height of the last Ice Age (about 20,000 years ago), sea levels were about 120 meters lower than they are today. The Bay was dry except for a few rivers, and the coast was about 60 km (dotted line) west of where it is today. As the Ice Age ended, sea levels rose again and covered the plain and filled the Bay.

effective than atlatl hunting. Thus, family groups could feed and defend themselves better, and villages became smaller and more dispersed. The effects of climate change are more difficult to ascertain. The area contains multiple microclimates that complicate matters, and these add to other events, such as coastline stabilization and estuary formation, to make interpretation difficult. In the late Holocene, a loss of resources caused the Native Americans to move from the coast to further inland (Warren, 2012). The arrival of the Spanish is one clear event, and it was signaled dramatic changes, including rapid population decline (possibly due to violence or disease) and changes to the environment.

During the last Ice Age, massive amounts of water were sequestered in the polar ice caps and in glaciers over much of North America. As sea levels dropped, a land bridge is thought to have appeared between Asia and Alaska. Humans walked across that bridge into North America. The date is unclear, but people seem to have moved into the New World in two or more waves, beginning around 16,500 years ago. They rapidly spread across the Americas. They arrived in the San Diego Bay Area over 10,000 years ago. By the time of contact with Europeans, Native Americans in California were estimated at 300,000.

When the first Native Americans arrived, sea levels were much lower than they are today, and the coast was 15 or more km west of its current location. The climate was temperate, and food was readily available. These first inhabitants were hunter-gatherers, and they did not till the soil. Thus, their effects on the environment were minimal, especially compared to those of the Europeans who arrived later. However, they were not without consequences.

The diet of the Native Americans in the late Holocene is unclear (Bartelink, 2009). Fish and other seafood were readily available, and these were a major part of the diet of the Coastal tribes. Plant foods were important, and overhunting might have played a role in the disappearance of large game animals. Climate change may also have been involved. The use of seafood also diminished (Bartelink, 2009).

The people of the La Jolla Complex are sometimes called the Shell Midden People. They depended on shellfish of various types, including venus clams (*Chione* spp.), scallops (*Argopecten aequisulcatus*), mussels (*Mytilus californianus*), and oysters (*Ostrea lurida*). Sea mammals and fish were also used at some times. The remnants in the middens are a key archaeological resource. They show what the people were eating and whether they were fishing for marine or littoral fish. These provide important insights into how they lived. Van Den Hazelkamp (2011) studied the middens in San Diego County. The dating of the middens is not complete, but some information can be gleaned by examining the vertical arrangement of the remains. He found some changes from rocky shore to bay to beach species in the remains. Those changes might indicate the silting up of the bays so that other sources came to be used, but the data are not completely clear. There does not seem to be any evidence that the people changed from eating shellfish to terrestrial game.

Native Americans lived around Batiquitos Lagoon at least 8,000 years ago and possibly longer (Batiquitos, nd). However, there is a gap in the archaeological record that ended about 1300 years ago. For unknown reasons, the area seems to have been deserted. Some speculate that climate change made the area unlivable. About 10,000 years ago, the last Ice Age ended, and the excess ice at the poles and in glaciers that had covered much of North America began to melt. Sea levels rose. Perhaps these changes affected the shellfish and fish that the Native Americans depended on. Eventually, about 3500 years ago, the sea level stopped rising, but the lagoon began to silt up and become more shallow. About 1300 years ago, people returned to the lagoon.

Several other groups were also represented. About 3000–7500 years ago, the La Jollan people either assimilated the original San Dieguito group or evolved from that group. Then 1000–2000 years ago, Yuman-speaking people arrived to assimilate the La Jollan group. They then split into two groups: the San Lois Rey and the Cuyamaca. About 1000 years ago, Shoshonean-speakers arrived.

By the time that the Spanish arrived in the 16th Century, there were about 20,000 Native Americans living in the San Diego Bay Area. At the time, five distinguishable American Indian groups were present in San Diego County: Luiseno, Cahuilla, Cupeno, Kumeyaay, and Northern Diegueño.

Effects on the Land

The Native Americans made few changes to the land. Yuman groups arrived from the east in about 1000 CE and became the Kumeyaay. They settled all across the area from Mission Bay to the Tijuana River Valley. The Kumeyaay lived in small family groups in semi-permanent villages along oceans and rivers.

Like most Native American populations, they lacked beasts of burden and the wheel. There is still some disagreement about the extent to which they used fire to clear land. Lightening is uncommon along the Coastal ranges of California, and so, natural fires are not sufficient to produce the landscape mosaics that are found in Southern California (Keeley, 2002). Those must have been produced by fires started by humans. Those fires converted the shrubland into grassland that better served the purposes of the Native Americans. By using this strategy, they prevented larger fires, renewed the land, and encouraged oaks to grow. The fires they set encouraged the growth of oaks, berries, and other foods and controlled insects. Although intense fires kill trees, the thick bark of oaks gives them a degree of protection (Long et al., 2017). Thus, the oaks survive firers better than conifers and bounce back more rapidly after the fire. Burning off the shrubland changed the environment and also increased the numbers of game and protected against predators and enemies.

The Kumeyaay were primarily hunter-gatherers. They hunted local game, such as rabbits, deer, quail (*Callipepla* sp.), mountain sheep (*Ovis canadensis*), antelope, fish, and shellfish. They did not conduct agriculture per se, but they did use forest gardening and manage and cultivate wild areas. They gathered prickly pear (*Opuntia* sp.), acorns (*Quercus* sp.), nuts, and agave hearts (*Agave* sp.) They also used seeds of grasses and forbs, legumes, mushrooms, and more. Native plants were used for much more than food. The provided medicine, body decoration, and art. The Kumeyaay used willow branches to ease pain. They ate cactus pulp as a source of water. mescal, agave, yucca, sunflowers, wild squash, and mesquite.

Acorns were a major item in the diets of most of the California Native Americans. An argument has been made that acorns did not become a stable until the 1700s. The Kumeyaay men, women, and children gathered acorns in the autumn. They ground the acorns into a flour (Hector et al., 2009) and then washed the flour extensively to remove the poisonous tannic acid. Acorns are rich in carbohydrates and fats, and they can be stored to provide food year round. Acorns are also stored by many animals, including bears, pigs, birds, squirrels, and rats. Acorns, berries, and other nuts can sustain a whole ecosystem. Black oaks (*Quercus kelloggii*) were a particular favorite for the Native Americans in Southern California (Anderson, 2005), but tanoak (*Notholithocarpus densiflorus*) and other types were also collected. Black oaks range from Oregon to Southern California. In addition to acorns, they provide homes for animals, including mule deer (*Odocoileus hemionus*), acorn woodpeckers (*Melanerpes formicivorus*), pileated woodpeckers (*Dryocopus pileatus*), mountain quail (*Oreortyx pictus*), and band-tailed pigeons (*Patagioenas fasciata*) (Long et al., 2017).

Oaks do well in these different environments. The Coast Live Oak (*Quercus agrifolia*) prefers the coastal plains. Blue Oaks (*Quercus douglasii* Hook) like the hot interior valleys. Black Oaks (*Q. kelloggii*) like higher elevations. The Interior Live Oak (*Quercus wislizeni*) seems to prefer riverbanks. The Valley Oaks (*Quercus lobata)* are the largest of the oaks, and they produce the largest acorns.

Early humans used what has been called "fire-stick farming," in which fire is used to clear ground, kill vermin, and regenerate plant food. It could open up woodlands so that more desirable plants could move in (Pausas and Keeley, 2009). Most archaeologists believe that Native Americans had stable foraging economies, and so, they did not need to develop agriculture (Bettinger, 2015). However, Berryman (2018) disagrees. He examined the sites on the Camp Pendleton Marine Base just north of San Diego and other records. His position is based on several observations. First, the climate in the last parts of the Late Holocene was much more stable than it had been earlier. The stability was much more conducive to establishing larger communities and nascent agriculture. In fact, there seems to have been a movement to larger and more communities at that time. He and others (Kimmerer and Lake, 2001) also cite the use of fire in the area by the Native Americans. Keeley (2002) estimated that 25% of the landscape had been burned off. Burning favors grasses over forbs. Next Berryman (2018) points out that the sites at Camp Pendleton show the use of grasses, legumes, and small edible seeds, which is similar to early agricultural sites in other areas in North America. No tools associated with agriculture have been found. However, the harvesting of seeds could easily have been done with baskets, and other tools might not have been needed.

Extinction of Large Mammals

About 50,000 years ago near the end of the Late Quaternary, many large mammals in North America and Eurasia began to be lost (Barnosky et al., 2004). The reasons are unclear but two causes have emerged as the most likely. North America had more extensive climate change and lost more mammals than Eurasia. The disappearance of the mammals coincided with the arrival of humans in those regions. Those animals included mammoths, mastodons, giant beavers, and others.

There is evidence for both models. Martin (1967), in his "overkill hypothesis," suggested that Native Americans hunted those mammals to extinction. Sandom et al. (2014) completed an analysis of all large mammals (greater than 10 Kg) that became extinct between the last Interglacial period (132,000 years ago) and the late Holocene (1000 years ago). They found that the loss of the animals was strongly tied to the arrival of humans. A correlation of the timing of 42 archaeological sites and genetic and climatic evidence also shows that humans arrived before the Last Glacial Maximum (26.5 to 19,000 years ago) but probably dispersed during a warming period (Greenland Interstadial) (Becerra-Valdivia and Higham, 2020). Smith et al. (2018) found similar evidence of human involvement. Still, not everyone agrees with the overkill hypothesis (Grayson and Meltzer, 2003). The evidence seems to indicate that only the mammoth and the mastodon were hunted. The rest were not. Also, temperatures fluctuated greatly near the end of the Pleistocene. The controversy is unlikely to be resolved soon, but Stuart (2015) strikes a middle ground by implicating both human and climate factors in the loss of the great mammals. Lorenzen et al. (2011) weighed the evidence by looking at the genetic diversity of the six large herbivores: woolly rhinoceros (*Coelodonta antiquitatis*), woolly mammoth

(*Mammuthus primigenius*), horse (wild *Equus ferus* and living domestic *Equus caballus*), reindeer/caribou (*Rangifer tarandus*), bison (*Bison priscus/Bison bison*) and musk ox (*Ovibos moschatus*). They concluded that climate was a major driver of the changes. It can probably explain the loss of the Eurasian musk ox and woolly rhinoceros. However, both climate and seem to be involved in the extinction of others, including Eurasian steppe bison and wild horse.

THE SPANISH

The first European to visit the San Diego Area was Juan Rodriquez Cabrillo in 1542. The next was decades later in 1602, when Sebastian Vizcaino arrived. Upon hearing that a good harbor just to the north, he dispatched Ensign Sebastian Melendes to explore what would later come to be known as Mission Bay. In 1769, the Spanish decided to establish a colony in Alta California to block any advances by the English and Russians on their territory. That same year, a group of Spanish established a camp near Old Town. They built a presidio near the mouth of the San Diego River. A couple of months later, Father Serra arrived and founded the mission San Diego de Alcala. They brought with them 200 head of cattle, the first in California (Larson-Praplan, 2014).

By 1823, the Spanish had founded 21 missions throughout the state. Their herds totaled more than 400,000 cattle, 61,000 horses, and 300,000 sheep. Examination of the animal remains at the San Diego Presidio showed 20 species of birds and 14 of mammals (Sasson and Arter S, 2020). Two breeds of chickens included a smaller bantam and a standard-size chicken. The types of bones found suggested that the chickens were used for meat and eggs. The Native Americans also started herding animals. Some of the animals escaped and started feral herds. While Native Americans used fire to clear land to some extent, the arrival of Euro-Americans (about 1850) significantly changed the incidence of fire in California (Van de Water and Safford, 2011).

The Spanish changed the world of the Native Americans. The area had great stands of oaks and willows. The Spanish seized the land, cut down the oaks to build houses. The early settlers burned off the native vegetation to clear land for grazing cattle and sheep and introduced cattle that ate the oak seedlings so that the older trees could not be replaced. Without the oaks and other native plants, the scare supply of water was reduced even further. The Spanish also introduced non-native plants and animals that further degraded the environment.

THE MEXICANS

In 1821, Mexico became independent of Spain, and Alta California, including San Diego, became a part of Mexico. At that time, about 600 people lived in the city. That number decreased into the 1830s. The new Mexican government began issuing land grants. The Spanish had not allowed private land ownership but had allowed individuals to have what were essentially free leases. The first rancho was Los Peñasquitos for 8,486 acres. Over time, they made 33 land grants that covered 2500 square kilometers. The Native Americans were pushed aside and, in many cases, used essentially as slaves. The lands were used mostly for grazing of cattle or sheep.

THE AMERICANS

The population of San Diego and the region has grown enormously since the arrival of the Americans. San Diego is now a major city with all of the development and infrastructure that such a city entails.

Mission Bay is an excellent example of how the environment has been changed. Mission Bay is one of the most modified bodies of water in southern California. In the last 150 years, the Bay was filled and dredged. Originally, Rose Creek, Tecolote Creek, and the San Diego River all flowed into False Bay. In 1853, a dike was built to change the course of the San Diego River so that it flows now into Mission Bay. This change prevented the discharge of large amounts of sediment into San Diego Bay where it was interfering with shipping. In the late 19th and early 20th centuries, Mission Bay was favored by fishermen and duck hunters. Over time, these sportsmen were displaced by development, and today, Mission Bay features large entertainment sites. The level of water exchange near the mouth is high, but it is much less in the back bay. An artificial island, Fiesta Island, also prevents flushing. Two creeks contribute organic-rich runoff to the Bay. The salinity is higher in the back bay region, probably due to evaporation of water. It is lower in the wet months.

San Diego grew slowly at first. According to the US census, its population was 650 in 1850 and 731 in 1860. The city's late connection to a transcontinental railway caused a population boom, and the city reached 16,159 by 1890. It grew very rapidly in the early part of the 20th Century, especially as the US Navy made major investments in the area. By 1950, the city had nearly 150,000 inhabitants, and in the 1980s, it passed 1 million. Such growth required an enormous amount of building housing, businesses, roads, sewers, waterlines, and more. The area around the Bay was changed significantly by all of these buildings. The major industries were the US Navy, aviation, and tuna fishing. All affected the quality of the water in the Bay.

Waste Management

Trash collection is always a challenge for cities, and San Diego had its share as the city grew in the late 19th Century. At that time, private contractors disposed of waste by burning it or by dumping it at sea. Some people simply dumped their waste in any vacant area, but that was no more acceptable then than it is now.

In 1908, the City hired H. L. Emerson to dispose of trash. Even in that day, they practiced a form of recycling. Food waste was separated out and used as hog feed. Also like today, people complained about the cost of hauling the waste away and tried to get out of it. Over time, the City came to be responsible for disposal. Individuals and institutions used burn dumps until the 1950s. The City opened landfills and passed ordinances to control the disposal of waste. Over time, they have tried various means to extend the life of the landfills, such as increasing the allowed height of the dumps. The County also opened landfills. The efforts to control waste continue.

Mining

The 1849 Gold Rush brought thousands of people to the San Francisco area hoping to strike it rich. Few did, but the hydraulic mining associated with the time recked environmental havoc and resulted in an enormous amount of sediment that filled rivers and streams and found its way to the San Francisco Bay.

In 1842, gold was discovered in the area around Julian. But the amounts were small, and the San Diego Bay region did not experience an onslaught like Northern California. However, the region it has had its own history with mining for gems and minerals. Pegmatite formations were found in the mountains near the Pala, Julian, Mesa Grande, and Pine Valley areas. Pegmatites are igneous rocks that form as magma is cooling, and they contain very large crystals as the magma cools slowly (King, nd). Most crystals are centimeters long, but some can be huge. For example, a crystal of spodumene (an ore of lithium) found in South Dakota was 13 meters long, 1.5 meters in diameter, and contained over 80 metric tons of spodumene. The rocks can be of variable composition, but most are similar to granite (e.g., quartz, feldspar, and mica). Some pegmatites contain the minerals lithium or beryllium. Others contain gemstones, such as tourmaline, aquamarine, and topaz. The pegmatites often at the margins of a batholith or pluton. In the pluton melt, water becomes trapped and superheated. It also collects high concentrations of various chemicals, and as the water escapes and is lost as steam, the chemicals remain to form large crystals (London, 2018). Much of San Diego County is part of two plutons, which are small parts of a larger batholith (see Chapter 2).

In the late 1800s, San Diego County was producing minerals and gems. Mines are fairly common: about 760 abandoned mines may still exist in the county. Among other items, the mines produced the gemstone tourmaline. Its 700 colors include dark rich rose to pinks and greens. From 1902 to 1910, San Diego provided imperial China with 110 metric tons of gem-quality pink tourmaline. That market dried up in the early 20th Century, but even today other gemstones are being found, including kunzite, morganite, topaz, and garnet.

POPULATION INCREASES

The first Californians were the Native Americans, and they lived in the San Diego Bay Area for thousands of years before the first Europeans appeared. The environmental impact of the Native Americans was limited. They built shell mounds, burned some areas for agriculture, and lived in relative harmony with the native animals and plants. Some anthropologists hypothesize that the arrival of humans in the Americas contributed to the extinction of many larger mammals. When the first Spanish explorers arrived, 200,000 to 500,000 Native

Americans were estimated to be living in California (Mosier, 2001). In 1542, about 20,000 Native Americans were in the San Diego area. Tragically, 90% of the Native Americans died after the Europeans arrived.

The number of Spanish colonists in California was always surprisingly low. At annexation to the US, foreigners outnumbered the Spanish in California by 9,000 to 7,500 (Gerston and Christensen, 2013). In 1838, the population of San Diego was reduced to 100–150 people. The population was 650 in 1850 and 731 in 1860, according to the U.S. census.

The military has been a large part of San Diego since the beginning. San Diego began as a Spanish military outpost, and the city's relationship with the military has grown dramatically. At the beginning of the 20th century, the civic leaders were seeking a way to increase commerce in San Diego (Martin, 2010). The city had enjoyed a "boom" in the 1880s, but more recently, business had slowed. They decided that they would try to convince the U.S. Navy to use their excellent harbor for a coaling station. This effort began a long and continuing relationship with the US Navy and US Marine Corps.

REFERENCES

Barnosky AD, Koch PL, Feranec RS, Wing SL, Shabel AB (2004) Assessing the causes of late Pleistocene extinctions on the continents. *Science* 306: 70–75.

Bartelink EJ (2009) Late Holocene dietary change in the San Francisco Bay area: stable isotope evidence for an expansion in diet breadth. *California Archaeology* 1: 227–252.

Batiquitos (nd) Native Americans. Living around Batiquitos. Batiquitos Lagoon Foundation. Retrieved from: http://www.batiquitosfoundation.org/wp-content/uploads/2011/06/Factsheet_Native_Americans.pdf; accessed February 13, 2021.

Becerra-Valdivia L, Higham T (2020) The timing and effect of the earliest human arrivals in North America. *Nature* 584: 93–97.

Berryman S (2018) What's happening in late Holocene Southern California, is it agriculture or what? *Southern California Anthropology Proceedings* 32: 36–50.

Bettinger RL (2015) *Orderly Anarchy: Sociopolitical Evolution in Aboriginal California.* University of California Press, pp. 137–138.

Braje TJ, Costello JG, Erlandson JM, Glassow MA, Johnson JR, Morris DP, Perry JE, Rick TC (2010) Channel Islands National Park Archaeological Overview and Assessment. Department of the Interior. Retrieved from: https://www.nps.gov/chis/learn/historyculture/upload/Final-Arch-Overview-May-2015.pdf; accessed February 13, 2021.

Braje TJ, Dillehay TD, Erlandson JM, Klein RG, Rick TC (2017) Finding the first Americans. *Science* 358: 592–594.

Erlandson JM, Moss ML, Des Lauriers M (2008) Life on the edge: Early maritime cultures of the Pacific coast of North America. *Quaternary Science Reviews* 27: 2232–2245.

Grayson DK, Meltzer DJ (2003) A requiem for North American overkill. *Journal of Archaeological Science* 30: 585–593.

Gerston L, Christensen T (2013) *California Politics and Government: A Practical Approach*, Cengage, Boston, p. 2.

Hector SM, Foster DG, Pollack LC, Fenenga GL (2009) An overview of Cuyamaca oval bedrock basin metates. *Proceedings of the Society for California Archaeology* 21: 161–168.

Keeley JE (2002) Native American impacts on fire regimes of the California coastal ranges. *Journal of Biogeography* 29: 303–320.

King HM (nd) Pegmatite. Geology.com. Retrieved from: https://geology.com/rocks/pegmatite.shtml#:~:text=Pegmatites%20form%20from%20waters%20that,%22pegmatite%20dikes%22%20are%20formed; accessed February 13, 2021.

Kimmerer FW, Lake FK (2001) The role of indigenous burning in land management. *Journal of Forestry* 99: 36–41.

Larson-Praplan S (2014) History of rangeland management in California. *Rangelands* 36: 11–17.

London D (2018) Ore-forming processes within granitic pegmatites. *Ore Geology Reviews* 101: 349–383.

Long JW, Goode RW, Gutteriez RJ, Lackey JJ, Anderson MK (2017) Managing California black oak for tribal ecocultural restoration. *Journal of Forestry* 115: 426–434.

Lorenzen ED et al., (2011) Species-specific responses of Late Quaternary megafauna to climate and humans. *Nature* 479: 359–364.

Martin J (2010) The San Diego Chamber of Commerce establishes the U.S. naval coal station, 1900–1912; San Diego's first permanent naval facility. *The Journal of San Diego History* 56: 217–232.

Martin PS (1967) Prehistoric overkill. In: *Pleistocene Extinctions: The Search for a Cause* (eds. Martin PS, Wright HE Jr.) Yale University Press, New Haven, pp. 75–120.

Mosier P (2001) A brief history of population growth in the Greater San Francisco Bay Region. In: *Geology and Natural History of the San Francisco Bay Area; A Field-Trip Guidebook* (eds. Stoffer PW, Gordon LC) US Geological Survey Bulletin 2188, chapter 9, pp. 181–186.

Pausas JG, Keeley JE (2009) A burning story: The role of fire in the history of life. *BioScience* 59: 593–601.

Sandom C, Faurby S, Sandel B, Svenning J-C (2014) Global late Quaternary megafauna extinctions linked to humans, not climate change. *Proceedings of the Royal Society B: Biological Sciences* 281: 20133254.

Sasson A, Arter S (2020) Earliest utilization of chicken in Upper California: The zooarchaeology of avian remains from the San Diego Royal Presidio. *American Antiquity* 85: 516–534.

Scharlotta I (2015) Determining temporal boundaries and land use patterns: hunter-gatherer spatiotemporal patterning in San Diego County. *California Archaeology* 7: 205–244.

Smith FA, Elliott Smith RE, Lyons SK, Payne JL (2018) Body size downgrading of mammals over the late Quaternary). *Science, 360*, 310–313.

Stuart AJ (2015) Late Quaternary megafaunal extinctions on the continents: A short review. *Geological Journal* 50: 338–363.

Van de Water KM, Safford HD (2011) A summary of fire frequency estimates for California vegetation before Euro-American settlement. *Fire Ecology* 7: 26–58.

Van Den Hazelkamp A (2011) Tides of change? Shell middens of the Mission Bay Area, San Diego, California. *Proceedings of the Society for California Archaeology* 25: 1–10.

Wade L (2017) On the trail of ancient mariners. *Science* 357: 542–545.

Warren C (2012) Environmental stress and subsistence intensification. *California Archeology* 4: 39–54.

Waters MR (2019) Late Pleistocene exploration and settlement of the Americas by modern humans. *Science* 365: eaat5447.

7 San Diego Bay Today

San Diego Bay was created over millions of years by the massive combined forces of tectonic plate movement, the rise and fall of ocean levels, volcanos, and wind and water erosion. The result was a beautiful natural bay and harbor surrounded by rich marshlands and beaches with an amazingly diverse array of plants and animals. Inland are mountains and deserts. Several rivers flow through the region. The forces that created the Bay work on a very long time scale, far longer than humans can easily appreciate.

However, the greatest changes to the Bay over the last hundreds of years were caused by humans.

San Diego Bay region covers a wide section of Southern California. It extends north to Del Mar and the Carmel Valley, south to Tijuana, and east to El Cajon, Otay Ranch, and Otay Open Space. It includes Mission Bay. San Diego Bay is 20 km long by 0.5–1.5 km wide. Several rivers either feed the Bay or are important in the Bay Area. Several rivers either cross the San Diego area, including the Santa Margarita, San Luis Rey, San Dieguito, San Diego, Sweetwater, Otay, and Tijuana Rivers.

The changes caused by humans have significantly changed the natural Bay, and they have occurred in the last few hundred years. The Bay itself has been dredged to maintain a clear shipping channel. Dredged material has been used to fill other areas. Most of the tidal flats have been filled for construction. Rivers were channeled and even redirected. Enormous amounts of building have taken place to make housing, employment, transportation, and entertainment for millions of residents. The water in the Bay and the air above have been polluted.

Here we will review the status of the physical Bay, including the land surrounding it, the water in the Bay and the rest of the region, the mudflats and marshlands between the land and water, and finally the climate and air about the region.

PEOPLE

The San Diego Metropolitan Area has a population of over 3.3 million. It is the 17th most populous area in the United States. The average increase in population for the rate of the city has been about 0.9% per year. That has slowed slightly in recent years, but it is still strong. The largest economic sectors are defense/military, tourism, and research/manufacturing. In 2014, the GDP of the region $206 billion. In 2012, San Diego welcomed 32 million visitors. Cruise ship stops hit their peak in 2008, when more than 200 ships visited the city. The region is deeply involved in trade. The port is a major entry and exit facility, and the

DOI: 10.1201/9780429487460-7

proximity to Tijuana makes San Diego a major crossing point for trade between the two nations.

The military footprint in San Diego County is simply massive. Over the years, the US Navy and US Marine Corps have constructed major facilities in the San Diego region. Today there are 16 installations in the county. These include Marine Corps Air Station Miramar (12,000 Marines), Marine Corps Base Camp Pendleton (42,000 Marines on 125,000 acres), Marine Corps Recruit Depot San Diego (trains 21,000 recruits each year), Naval Base Coronado (5000 military and 7000 students and reservists), Naval Base Point Loma (22,000 personnel), Navy Base San Diego (20,000 military and 6000 civilians), and US Coast Guard Station San Diego. These installations include large numbers of active duty servicemembers along with a substantial civilian workforce and families. San Diego is homeport to many US Navy ships. These include 14 amphibious assault ships, eight cruisers, 16 destroyers, 13 littoral combat ships three mine-countermeasure ships, and five supply and support ships.

Finally, all of the people who work at these facilities require housing, transportation, entertainment, water, sewage, schools, and more. In addition, the military needs space for training and maintenance of numerous ships and aircraft. All of these categories stress the natural environment of the San Diego Bay region.

LAND

The land includes mountains, low lands, mudflats, and seashore. The basic structure of the Bay region has remained the same for a few million years. The tectonic forces that formed the structure act too slowly for human eyes, and the volcanoes that were powered by the subduction of the Farallon plate and slab windows (Wilson et al., 2005) have long gone extinct. More recent changes have been due to water, weather, and humans.

Earthquakes and Other Movements Caused by Plate Movements

Earthquakes are an ongoing reminder of the titanic forces that built California (see Chapter 2). The San Diego Bay Area and the rest of California receive those reminders periodically. Fortunately, most of the earthquakes are very mild, but some are severe and cause death and destruction. The San Diego region has experienced several strong earthquakes in the last centuries, including 1800 (magnitude: 6.5), 1862 (m6.0), 1804 (m5.75), 1986 (m5.4), and 1986 (m4.7). Everyone in California knows that the next "Big One" is certain to come at some point. In fact, over the tens of millions of years, there have been thousands of Big Ones.

For San Diego, the most serious threat is thought to be the Rose Canyon Fault, which crosses through the most populated areas of the region from La Jolla in the north to Tijuana in the south (EERI, 2020). There is an 18% chance of a magnitude or larger earthquake on this fault in the next 30 years (Field et al., 2015).

An earthquake of this magnitude would result in major damage to the infrastructure of the region. Movement at the surface could be as much as 2 meters. The damage would include surface fault rupture with severe shaking. Fires are also a major risk after a powerful earthquake. The San Diego Bay Area also has several other significant faults, including the Elsinore and San Jacinto faults. Even a major earthquake on San Andreas fault to the east could cause considerable damage in San Diego. Any of these could be a problem.

Other movements have resulted from the actions of plate tectonics. For example, the mountains to the east of the Bay continue to rise as they are pushed up by the pressure accumulated from the bend in the San Andreas fault in Southern California. These mountains are the results of the folding of the crust.

Other Land Movement

Earthquakes are not the only ways that soil and rock move. Mud and silt flow downhill in a manner similar to water. Landslides result from excessive rainfall, earthquakes, and other factors. Mud and other debris flows, and a heavy rain accelerates the actions. The average rainfall is greater in the northern part of the state, and so the landslide threat is greater there, but landslides also occur in the drier southern regions.

The signs of a pending landslide include ground cracks, sinking areas, leaks in pipes, tilting poles or fences, new cracks in structures. Rock strength and slope can be combined to create classes of landslide susceptibility (Wilson and Keefer, 1985; Ponti et al., 2008; Wills et al., 2011). Using these measurements, slopes can be rated. On low slopes, landslides are less frequent regardless of the strength of the rocks. The risk increases with the slope and the weakness of the rocks. For thousands of years, humans have been a major factor in earth movements. Clearing land for crops, housing, and other developments encourage erosion that fills rivers and streams. Deforestation also encourages erosion, and dams have stopped the natural movement of the material (Voosen, 2020). Landslides cost lives and damage property.

New space-based imaging technologies, such as synthetic aperture radar interferometry (InSAR), have provided significant improvements in measuring surface deformations that can be informative to examining faults, ground movements up and down, landslides, and more. Uplift results from the movement of the plates along the faults. Downward movements can result from depletion of aquifers and settling of unconsolidated sediments.

Slow-moving landslides are a problem throughout much of California. They rarely cost lives, but they can be extremely destructive to structures and infrastructure. The slides depend on the soil, climate, and earthquake activity of the area, but the mechanisms that initiate and maintain them are not well understood. They often occur in soils rich in clay and rock that are mechanically weak and have high levels of seasonal precipitation. Cohen-Waeber et al. (2013) used a network of global positioning stations (GPS) and InSAR to measure surface displacements with extreme accuracy. The area that they examined is part of the

California Coast Range, which is a mix of Jurassic to Tertiary sedimentary, volcanic and metamorphic rocks. It is overlain with Quaternary colluvial and alluvial deposits. Importantly, this soft rock has been subjected to considerable fracturing and weathering and is susceptible to sliding. They found that the land movements and velocities were quite similar. They found only limited influence of earthquakes, even though some small to moderate earthquakes occurred during the measurements. Rainfall seemed to be a key element in the initiation of landslides.

The California Coast Ranges are an ideal place to study slow landslides. Lacroix et al. (2020) reviewed the forces that control slow landslides in this environment. Those factors include the overall geology, climate, and tectonics and also precipitation and groundwater, earthquakes, river erosion, anthropogenic activities, and external material supply. Rivers can block slow-moving landslides, and landslides can block rivers. Finnegan et al. (2019) examined the several effects on the blockage of rivers by landslides and found that wider rivers are less affected by landslides. More narrow streams can be completely blocked by the landslide. Also, rivers vary in their ability to mobilize the material in the landslide.

Since landslides are often associated with extreme weather events, Cordeira et al. (2019) compared landslides from 1871 to 2012 with records of Pacific winter storms and atmospheric river events in Northern California. They found that 76% of the landslides occurred during storms and 82% occurred during an atmospheric river.

Landslides do not just occur on dry land. They happen at sea as well, and those landslides can cause extreme damage through tsunamis. Underwater landslides can be related to slides, slumps, reef failures, and earthquakes. Not all produce a tsunami, but some do. Local tsunamis are dangerous because there is little time for a warning (Please see the section on tsunamis later in this chapter).

Coastal Erosion

The bluffs along the San Diego coast are sandstone, a relatively soft stone that weathers easily. Those cliffs will likely be under increased pressure as climate change increases the number and severity of storms in the area. These effects will make planning even more important. Complicating matters is a report showing that pass erosion on cliffs is not an effective predictor of future behavior (Young, 2018).

Coastal erosion is a serious problem in many areas (Figure 7.1). The cliffs can be 5–115 meters high, and as the population has grown, so has the number of structures on the edge of the cliff. In addition to houses on the cliff edge, various components of infrastructure, such as power lines and roads, are also located there. For example, near Del Mar, railroad tracks run along the ocean cliffs and face the risk of collapse into the sea. In fact, six major and over 100 minor collapses have occurred since the summer of 2018 (Mulkern, 2019). The rail lines were put there because people believed that the cliff face was stable, but it is not at all.

FIGURE 7.1 Coastal erosion. The cliff face is eroding under the pressure of waves striking the base of the cliff near Isla Vista. This type of erosion threatens much of California's coast. (Photograph courtesy of Alex Snyder, US Geological Survey).

As the cliff top retreats in the face of wave action, the structures and land on the cliff top are threatened with collapse into the sea. This type of erosion has been going on for millions of years, and so, protecting these exposed areas is very difficult. Various measures have been used to slow the inevitable, including beach replenishment, emplacement of boulders to absorb the wave action, and retaining walls as high as the cliffs (Young and Ashford, 2006). As a result, nearly 10% of the coast of California has some form of protection. Young and Ashford examined the effectiveness of those protections. Overall, the efforts slowed the cliff retreat by 43%, compared to those adjacent areas that lacked protection. Measures that protected both the upper and lower aspects of the cliff fared better than those that only protected one or the other.

The actual erosion processes are quite complicated and require careful measurements to understand what is happening in a particular area. Johnstone et al. (2016) used terrestrial laser scanning and sediment analyses to construct models of cliff erosion. They repeated their analysis at 12 sites along 20 km of coastline every year for 6 years beginning in 2006. A number of factors are involved, including the rocks that make up the cliff, beach width, elevation, precipitation, and other factors. They found that erosion on wide beaches is dominated by subaerial factors and on narrower beaches, it is comminated by waves. Their findings provide a baseline for future studies of coastal erosion. Young et al. (2010) used two relatively new technologies to examine a 400-m length of coast in Del Mar, California. They found that the two systems correlated well with each other, but the terrestrial lidar was much more accurate than the airborne system for detecting small changes. However, the airborne system could more rapidly examine cliffs and thus has its own advantages.

Another key factor in beach erosion involves the energy of the waves striking the beach. Benumof et al. (2000) studied erosion on the coast in San Diego County by looking at the height, energy, and power of waves hitting the cliff face. The assumption has been that waves eat into the cliff at the base. They found that the waves are not as important in cliff erosion as the material properties of the cliffs themselves.

Waste Disposal

More than 3.5 million people live in the metropolitan area, and that many people create a lot of waste that must be properly disposed of. Unfortunately, San Diego County is losing its battle with trash. In 2017, each person in the county created 2.5 kg of waste per day. In 2018, that amount had grown to 2.6 kg (Equinox, nd). The waste numbers are actually down from a high in 2005 of just under 3.6 kg per person per day.

Landfills are much improved over what they were years ago. Today they seek to minimize air pollutants and water leaching compounds as it percolates through the material. Nevertheless, the degradation of organic waste in the landfill creates methane, a greenhouse gas. Landfills and composting facilities produce 15–20% of the state's methane emissions. Landfills also require a lot of space. In 2015, San Diego County had enough landfill space for 110 million metric tons of trash, and several other sites were under development. In 2019, the city planned to increase the capacity of the Miramar landfill by increasing its height by 7.5 meters (Garrick, 2019). There is no more room to expand horizontally. Trash will now be piled 155 meters. In 2020, there are six active landfills in the county. The city has an active recycling program, but still, about 90,000 metric tons (10–15%) of the recycled material is sent to the landfill each year (Wood, 2019).

Waste disposal in the 1950s was much less careful than it is today in San Diego (DeWyze, 2000). In most of those years, the city ran a 115-acre landfill in Mission Bay Park on the southern shore between SeaWorld and Interstate 5 near Fiesta Island. It accepted most anything. These included millions of liters of industrial waste from the local aerospace industry. Some was in steel drums. Some was simply poured into holes in the ground.

That landfill was closed in 1959, but there has been no clean-up of the site since. In fact, no one knows what is actually buried there. There were no laws regarding hazardous waste disposal until the 1970s. Speculation of what was dumped includes chromic, hydrofluoric, nitric, sulfuric, and hydrochloric acids, alkaline solutions, and paint and other wastes. Later reports added carbon tetrachloride to the list. The site was covered with material dredged from Mission Bay.

Another report from the 1980s found more than 60 pollutants described by the Environmental Protection Agency as "priority," including 12 heavy metals, 38 organics, and 12 pesticides.

An effort to build a boating facility called South Shores Park in the area led to the release of several chemicals, such as hydrogen sulfide, dichloroethane, trichlorethane, and carbon tetrachloride (Martinez, 2006). The fish and other aquatic animals have concentrated the pollutants, and may not be safe to eat.

A new study of the site in 2002 stated that it is not an environmental threat now, but has some possibility of becoming one in the future. Opinions differ on a solution. Some want to move the material to a better site. Others contend that disturbing the site would be a cure worse than the disease.

In 2015, Sea World hoped to build a new hotel and resort center on the site, but the city prevented it (Gormlie, 2015). A three-story hotel would have required considerable excavation that might have released toxins. In 2018, Sea World scrapped the project.

Another abandoned dump occupies a 51-acre site once used by the US Navy. It is near the airport, unlined, and only 150 meters from San Diego Bay. There is a large-scale fuel plume in Point Loma.

These examples are in San Diego, but our intent is not to malign the city. Sadly, the practice of dumping wastes without regard to safety was common throughout the United State.

WATER

The greatest source of water to the Bay is the Pacific Ocean, and the water in the Bay moves rhythmically with the waxing and waning of the tides. Freshwater enters from precipitation, outflows of rivers and creeks, and discharge from wastewater treatment plants.

San Diego Bay is a graben, which is a structural depression due to subsidence between two faults. The result is a shallow estuary. Its depth averages about 12 meters and it varies from 18 meters at the mouth to less than 1 meter at the southern end. Before all the dredging and filling, it was more shallow and wider. Because the Bay is open to the ocean, it is tidal. High and low tides differ by about 2.9 meters and are the strongest near the mouth of the Bay. In a single tidal cycle, about 30% of the volume of the Bay is exchanged. This tidal prism amounts to about 74 million cubic meters. The flushing varies between the mouth and the deepest part of the Bay. At the mouth, a complete flushing occurs about every 1–2 days, and at the southern end, it takes about 7–14 days.

Unfortunately, particularly over the last couple of centuries, the Bay has been a dumping ground and disposal site for many human activities. As a result, the quality of the water has been heavily impacted. Fortunately, in more recent years, efforts to improve the environment have achieved a number of successes. Those improvements are described in some detail in Chapter 9.

Freshwater

Freshwater enters the Bay from the Sweetwater and Otay Rivers and several smaller streams. Several others flow near the Bay and are important in the region, including the Santa Margarita, San Luis Rey, San Dieguito, San Diego, and Tijuana Rivers.

The minerals and microorganisms in freshwater reflect the rocks that it flows through on its way to the San Diego Bay (Robbins et al., 2018). For example, at the source of the San Diego River, the rocks contain iron and sulfate in the form of pyrites. As the river continues, water containing sulfates and magnesium oxides enters from other sources. Those minerals and the environmental conditions along the river support the growth of different microorganisms that, in turn, change the oxidation state of the minerals and cause the precipitation of some.

The course of the San Diego River has varied over the last 200 years. Originally, it flowed directly into San Diego Bay. In 1821, a sudden flood changed its course into Mission Bay for some years, but later it returned to its earlier banks and San Diego Bay. However, the river also brought a lot of sediment that began to interfere with shipping in the harbor. In 1853, the federal government built a levee that directed the river to Mission Bay. It was destroyed in a flood that same year but was rebuilt in 1876. Since then, the river outflow has remained in Mission Bay. El Capitan Dam has since reduced the amount of sediment entering Mission Bay.

Ground Water

The San Diego Formation Basin is a large confined shallow aquifer that lies under Imperial Beach, Chula Vista, National City, and the southern parts of San Diego. It contains about 960,000 acre-feet of water, and it begins only 33 meters under the ground. Obviously, it is near to the Pacific Ocean, and its water is brackish. Two desalination plants produce drinking water to reduce the amount imported from the Colorado River.

Because the aquifer is so close to the Ocean, intrusion by seawater is a concern that limits the use of the aquifer as a source of water. Anders et al. (2014) used geochemical methods to study the aquifer. They tested for levels of chloride and isotope ratios (e.g., strontium, deuterium, oxygen 18, and carbon 14). They found that groundwater movement in the aquifer results mainly from three actions. First, local precipitation recharges the shallower parts of the system. Second, precipitation in the mountains recharges the deeper paths of the saline aquifer. Finally, not surprisingly, intrusion of seawater from long ago. The chemical composition varied with depth. The concentrations of chloride decreased in the deeper portions of the aquifer, suggesting that most of the recharge came through that path.

Removing water from an aquifer runs the risk of encouraging subsidence. To better understand the relationship of groundwater removal and subsidence in the San Diego region, Brandt et al. (2020) explored change in ground levels with a powerful new technology called Interferometric Synthetic Aperture Radar. They examined ground levels before and after groundwater was removed and as it was replenished. They found that ground levels dropped about 75 mm upon removal, and that about 45 mm of uplift occurred after a period of recovery. The numbers are somewhat encouraging, but more study is needed to determine if the groundwater can be pumped out without significant subsidence.

Bay Water

Sediments

Sediments occur in nearly all bodies of water, including San Diego Bay, and they are critical to a wide variety of plants and animals. Invertebrates burrow and feed in the sediment. Fish eat those invertebrates or the plants growing there, and the progression goes on up the food chain. The sediments are affected by the kind of sediments that are washed into the Bay and by the waves that continually move them and the terrain under the water. Human activities also affect the sediments. Mining, agricultural, construction, and other work can increase the amounts of sediment or change its character. Dredging disturbs the sediments. In addition, sediments can bind and transport toxin and other pollutants.

Two areas of particular interest were examined. Latker et al. (2020) characterized the sediment grain size and chemistry in the waters around Point Loma Ocean Outfall. Chadwick et al. (1999) examined sediment near Naval Station San Diego, which had been a source of considerable pollution. They found that some pollutants (e.g., copper, mercury, and zinc) were elevated, and others were sometimes found at elevated levels (e.g., silver, lead, PAHs, and PCBs). Levels of arsenic, chromium, and nickel were mostly below target thresholds. Biological effects were examined with bioassays, benthic community structure, in-situ biomarker assays, and bioaccumulation studies. They found it hard to correlate measurements between the chemical contaminants and biological effects.

Fairey et al. (1996) examined sediments from 350 sites (e.g., Bay, Mission Bay, San Diego River, and Tijuana River) from 1992 to 1994. In the samples, copper, mercury, zinc, total chlordane, total PCBs and the PAHs were higher than desirable. They assessed habitats for diversity, species composition, indicator species, and more. They compared their results to established guidelines and noted that toxicity is likely to be a summation of the contributions of multiple compounds. Seven of the testing stations were determined to be the most polluted. They were in the Seventh Street channel area, two navy shipyards near the Coronado Bridge, and the Downtown Anchorage area. These results will aid the efforts to clean up the most affected areas.

Water Characteristics

The physical and chemical characteristics of the water are critical for the plant and animal life in the Bay. These include water temperature, salinity, and levels of various nutrients, such as phosphate, nitrogen, carbon, and organic carbon.

These might also change with the season or at different depths. In addition, the location of the sampling—at the mouth of the Bay where the influence of the Pacific Ocean would be keenly felt or deep inside the Bay where the water is farther from the Ocean—might result in differences.

Delgadillo-Hinojosa et al. (2008) assessed many of these characteristics throughout the Bay. Their measurements were taken at four times distributed across the summer and winter seasons. In the summer, they found that the temperature, salinity, and other measurements were higher in the outer bay than in the open ocean and higher still at the southern end of the Bay. They attributed the hypersalinity to evaporation during the summer. The winter readings were more interesting. Phosphate and organic carbon levels were higher near the mouth of the Bay, but salinity levels decreased the farther south they went. This result was likely due to the runoff of freshwater into the more southern end of the Bay. The mixing of some of the other compounds seemed to be more complex.

The temperature of the Bay can also be influenced by human activities. Many power plants located next to large bodies of water use that water to remove excess heat from electrical power production. The water runs through the plant one time and is returned to the bay, ocean, or river that it came from. This is an efficient, but now antiquated, method for cooling. The problem is that significantly warms the source of the water with detrimental effects on the plants and animals living there. However, in the case of green turtles (*Chelonia mydas*), an endangered species in San Diego Bay, the warm effluent attracts the turtles and makes them more active. In the summer, the turtles leave the area when the water temperature gets too hot. Turner-Tomaszewics and Seminoff (2012) took advantage of this observation to study the effects of temperature in the Bay on the turtles.

While studying fish types and numbers, Allen et al. (2002) recorded various physical characteristics of the Bay over 5 years. They found that the water temperature varied from a low of 14.0°C in January 1995 and 1997 to a high of 27.3°C in July 1997. The temperatures also varied by location: Those near the mouth of the Bay were colder by 2–5°C than those at the southern end. Surface salinity was relatively stable and only varied from 39.8 to 33.4 ppt. The stability was maintained even in years with heavy rainfall in January. Salinities varied. In one year, they increased until October and then fell off in January. In another year, they were higher in summer.

Pollution

For the overwhelming majority of its existence, San Diego Bay was a shallow bay surrounded by salt marshes and mud flats. Early human settlers built towns and filled some of the Bay, but their effects were small compared to those that began and accelerated in the 20th century. Construction work soared so that most of the bay shore is now occupied by buildings, wharfs, and military and industrial works. A shipping channel was dredged, and most of the mudflats and salt marshes were filled and drained. By 1950, the entire Bay was polluted with domestic and industrial waste. Visibility in the water was less than 1 meter, oxygen levels were half of normal, and fish had almost disappeared.

San Diego Bay receives waters from urbanized, military, and industrialized areas. Unfortunately, the Bay is one of the most contaminated areas in the United States (Bay et al., 2016). The pollution is in the water and in the Bay sediments, and it comes from the military, the shipyards, industry, power production, and urban run-off. The pollutants are polycyclic aromatic hydrocarbons, polychlorinated biphenyls, Tributylin (a component of anti-fouling paint), and heavy metals (e.g., arsenic, copper, lead, mercury, and zinc), and Diazinon (an insecticide).

Sewage

In 1888, a 40-mile sewage system was completed. It included an outfall in the Bay off Market Street. Completion of this system started the decline of the quality of the water in the Bay. For much of the 20th century, San Diego Bay was polluted by raw sewage. As the cities surrounding the Bay grew rapidly, no effective system was built to handle the problem of human waste, and as a result, millions of liters of sewage went directly into the Bay. Fortunately, in 1963, the Point Loma Treatment plant came on line, and the Bay began to recover. That facility has been continually ungraded over the years. Now sewage is not a serious problem for San Diego Bay (Delgadillo-Hinojosa et al., 2008).

The Point Loma Wastewater Treatment Plant is a highly efficient facility. Its maximum capacity is 910 million liters each day, and it typically treats 550 million liters each day. Solid removal is 88–90%, and the biological oxygen demand removal rate is 60%. Their processes involve screenings to remove large solids (e.g., plastic and wood), and then, iron chloride adds a positive charge to the solids. By bubbling air through the material, the surface tension is broken, and the inorganic material can settle. The positively charged solids are treated with an anionic polymer so that they settle out. The water is then heated anaerobically to 98.6°F. The methane released by the heating is burned to provide energy for the plant. The iron salts are recovered by adding hydrogen peroxide. They also use wet scrubbing and carbon polishing, and the resulting solids are transported to landfills. Treated waste water piped 7.25 km offshore and released into the Pacific Ocean.

One of the most serious recent releases of raw sewage occurred in 2006. A construction crew on the Nave base mistakenly connected a sewage line to a storm drain. About 53 million liters of raw sewage flowed into the Bay via Chollas Creek over 2 years until the mistake was discovered and corrected.

Although it is not directly connected to San Diego Bay, the region has another significant problem with sewage. The Tijuana River begins in Mexico, crosses the border, flows another 6 km, and empties into the Pacific Ocean. The last 1.2 km are a broad mud flat estuary called the Tijuana River Estuary. It is home to more than 370 species of birds. The river only flows after heavy rains. For most of the time, all of its water is used in Mexico. Unfortunately, the heavy rains can also result in flooding and sewage runoff. In 1944, the two countries signed the United States-Mexico Water Treaty, which covered the Colorado and Tijuana

Rivers and the Rio Grande. Mexico completed their flood control measures, but the US did not. So the flooding and sewage remain problems. The outflow of partially or untreated sewage is carried out into the Pacific, but it often causes beaches to be closed for kilometers. Sometimes the closures reach as far north as Coronado.

Over the decades, many attempts have been made to solve this very serious problem, but unfortunately, no fix has been able to keep up with the astounding population growth in Tijuana, and the River is still one of the most polluted. Raw and undertreated sewage is only one challenge. The water also contains DDT, hexavalent chromium, pathogens, and carcinogens.

Polycyclic Aromatic Hydrocarbons

Polycyclic aromatic hydrocarbons (PAHs) are a significant contaminant of San Diego Bay. Some occur naturally, but most are the result of human activities. They are produced when coal, oil, wood, garbage, or other materials are burned. Some PAHs are generated commercially in chemical production (e.g., naphthalene). Some are considered to be carcinogenic. Since they are the products of burning, they tend to vary greatly in size and composition. Many are insoluble in water, but they do bind to particles and so can be contained in sediments in the water.

Neira et al. (2017; 2018) examined PAHs in the sediments in San Diego Bay. Specifically, they looked at three recreational marinas and focused on specific areas where stormwater entered the Bay. Recreational boating is very popular and has grown rapidly. Many activities there might lead to spills of chemicals into the Bay, such as fueling, engine repairs, leaching from creosote-treated piles, and stormwater runoff. Marinas tend to be somewhat enclosed, and so, they trap contaminants in their sediments. Thus, marinas are hotspots for pollution. Neira et al. noted significant differences among the marinas. The most common species were those with four to six rings, and the levels of potentially carcinogenic molecules was high in all samples. They concluded that most of the PAHs come from aerial deposition and stormwater drainage.

Polychlorinated Biphenyls

Polychlorinated biphenyls (PCBs) are oily clear or yellow liquids or solids. Because they are resistant to extreme temperatures and pressures, they were widely used in electrical equipment, hydraulic fluids, lubricants, plasticizers, and more. Although their manufacture was banned in 1979 by the US Environmental Protection Agency, they remain in the environment from earlier leaks, spills, and improper disposal. PCB levels are declining since no more are being made, but they are also persistent. PCBs collect in sediments and can accumulate in fish and shellfish. Neira et al. (2018) examined levels of PCBs in surface sediments in three marinas in the Bay (i.e., Shelter Island Yacht Basin, Harbor Island West, and Harbor Island East). The levels differed, and the heavier groups of PCBs were predominant. Those at Harbor Island East were the highest.

Pollution has wide-ranging effects on plants and animals. The classic example is the thinning of bird egg shells by DDT. The thin shells break and allow the infant birds to die. The U.S. Fish and Wildlife Service sponsored a 2-year study of the effects of pollutants on birds and fish at the South Bay Salt Works in the San Diego Bay National Wildlife Refuge (Zeeman et al., 2008). Failed egg shells of various species were examined for thickness and presence of pollutants, such as metals, organochlorine pesticides, organotins, polychlorinated biphenyl (PCB) congeners, and polybrominated diphenyl ethers (PBDEs). The last two are related to DDT. The birds examined were black skimmers, Caspian terns, elegant terns, and least tern. The fish were California killifish, topsmelts, and longjaw mudsuckers. The egg shells were thinner than normal. The eggs had high concentration of DDE and PCBs. The concentrations in the fish samples were much lower.

Metals

Copper is a common contaminant and toxic to marine organisms, and its levels in estuaries are of interest. It comes from several sources. Copper leaches out of antifouling hull paint and is discharged from industrial plants and stormwater, and is released when sediments are resuspended (e.g., ship movements, dredging). Copper exists in seawater in multiple forms, including in dissolved complexes, colloids, and particles. However, the most toxic form is thought to be the free hydrated ion. In this form, it is chelated by a number of organic molecules.

Blake et al. (2004) determined the levels of copper at various location in San Diego Bay over about 13 months. They found the levels of total copper were lower at the mouth of the Bay (8 nM) and higher deeper in the Bay (55 nM). However, levels of free copper were just the opposite. The highest levels were found just after a period of rain in the winter.

Pollution affects many plants and animals, but long-lived animals feel the effects even more. The green turtle (*Chelonia mydas*) is an endangered animal in San Diego Bay. The development of the Bay region limits their critical environment and perhaps their health. Komoroske et al. (2011) compared turtle weights to levels of pollutants in their body. The amounts of metals (except mercury) in the carapace and 4,40-dichlorodiphenyldichloroethylene and chlordane in the blood were higher than those in other sea turtle studies. Levels of mercury were particularly worrisome because they exceeded the levels thought to interfere with immunological defenses.

San Diego Bay is contaminated by metals. Deheyn and Latz (2006) studied 15 elements in the sediments of the Bay and found that metal concentrations were higher at the mouth near the Ocean than deeper in the Bay. They also determined the bioavailability of those metals in brittlestars (*Ophiothrix spiculata*) from outside the Bay. They found that metals were concentrated at the bottom of the Bay in the sediments.

Thompson et al. (2009) determined the pollutants associated with sediments at several sites in the Bay. Only 10 contaminants were examined, although many other contaminants probably there may also affect organisms. They suggested that there was no right or wrong answers to the contaminants. It will ultimately depend on negotiations between regulators and stakeholders.

Trace metals can also be a problem for living organisms. For example, green sea turtles (*Chelonia mydas*) are affected by human activities in the San Diego Bay. Komoroske et al. (2012) examined trace metal contamination (e.g., silver, cadmium, manganese, selenium, and zinc) in the food of the turtles. Copper and manganese were found in a gradient with the greatest concentrations near the mouth of the Bay. Other trace metals were found in different amounts in various parts of the Bay. Understanding how the metals are distributed will help in understanding habitat utilization for the turtles.

Foraging green sea turtles were followed at the Seal Beach National Wildlife Refuge and San Diego Bay. Both are near urbanized areas. Barraza et al. (2019) assessed 21 trace metals in the scute (the back of the shell) and red blood cells. The contaminants were specific for each location. The Seal Beach turtles had more cadmium and selenium in red blood cells and more selenium in scute samples. Turtles from these two locations had higher trace metal concentrations than those in non-urbanized areas.

Plastics

Plastics are a major type of pollution (Auta et al., 2017). They can be in the form of microparticles or small pieces of plastic. Van et al. (2012) collected plastics from beaches around San Diego. They found plastic pellets and larger pieces that also contained PAHs, PCBs, DDT, and chlordanes on the plastics. Storms and work on the beach were key factors in determining the amount and types of plastics found.

Plastic debris raises many concerns. For example, it can completely cover coral reefs and entangle animals and animals can ingest the plastic. Chemical pollutants, such as insecticides, toxins, carcinogens bind to plastics. Metals also bind to the plastic and accumulate. Rochman et al. (2014) examined levels of chromium, manganese, iron, cobalt, nickel, zinc, cadmium, and lead that had bound to five types of plastics (e.g., polyethylene terephthalate, high-density polyethylene, polyvinyl chloride, low-density polyethylene, and polypropylene) at three sites in San Diego Bay. They found that all of the plastics accumulated all of the metals over time although high-density polyethylene had the lowest levels. Chromium, manganese, cobalt, nickel, zinc, and lead did not reach saturation on at least one type of plastic each, and that observation suggests that the plastics can accumulate very high levels of metals.

Microplastics

Microplastics smaller than 5 mm are a serious problem in marine environments and estuaries (Masura et al., 2015). Microplastics result from many everyday products and processes, including personal care products, plastic products, fabrics and fishing line, photodegradation of plastic items, cigarette filters, and more. Worse still, many personal care products contain microbeads, and billions of these particles are released into river, streams, and bays every day in the US. For example, large amounts of microplastics are released during the mechanical and chemical actions of washing synthetic fabrics. De Falso et al. (2019)

examined the effluent of household washing machines and found 124–308 mg per kg of fabric (640,000–1,500,000 particles). Microplastics are produced on land, but are washed into the Bay by run-off and water treatment plants. Water plants actually produce only small concentrations, but they also process so much water than the total number of microplastics released is huge. It can be millions of particles every day (Mason et al., 2016). Even a polystyrene cup lip can produce nearly 108 nanoparticles after 56 days in a weathering chamber (Lambert and Wagner, 2016). The particles are so small that they easily pass through the filters in wastewater treatment plants and eventually get into the food chain. Although small, they can physically block the digestive tract.

More importantly, they carry bound organic pollutants, including PAHs and heavy metals. Brennecke et al. (2016) showed how heavy metals adhere to plastics. The combined antifouling paint with polystyrene beads and polyvinyl chloride fragments in seawater. The seawater leached copper and zinc from the paint that was absorbed by both plastics in significant amounts. The amounts of copper in some marinas can be high, and that has led some to suggest that copper-based antifouling paints should be phased out in favor of better alternatives (Carson et al., 2009).

Toxicity in Specific Areas

Several watersheds feed the Bay, and each has its own characteristics and challenges (SDState, nd).

Los Peñasquitos Watershed

From 1966–2000, the watershed was rapidly developed (White and Greer, 2006). The urban land use increased from 9% to 37%. The character of the creek was changed by the addition of new structures and paved surfaces. Development also brought water from the municipal system. The vegetation along the river also changed. All of these factors resulted in much more runoff during the dry season (summers). The levels of dissolved solids and fecal-associated bacteria are high. The water from Los Peñasquitos Creek eventually empties into the Pacific Ocean at Torrey Pines State Beach north of San Diego. Unfortunately, the levels of fecal coliform bacteria are so high that they result in the closing of the beach. This occurs especially after a rainfall. The runoff enters the stormwater system and flows to the ocean. It can be dispersed many kilometers north and south of the outlet. A number of factors control the movement of the discharge plume, including the shape of the beach, wind strength and direction, tides, waves, and rain, and time (Hea and He, 2008). The urbanization of the creek area also affected the Lagoon that is kilometers downstream near the outlet to the Pacific Ocean. Greer and Stow used aerial photographs of the area taken in 1928 and 1999 along with other data to explore the changes. About 80% of the vegetation had changed, and the changes depended on the vegetation type. There was more marsh and riparian vegetation and less salt panne and mudflats. In addition, the summer stream discharge was 10-fold greater.

San Luis Rey Watershed

This watershed is north of San Diego. It is a mixture of open space, agricultural (cattle grazing, nurseries, citrus, and avocado groves), and residential areas. For much of the year, there is little water in the river. It flows freely in the winter months. Urban growth, mining, and agriculture have all contributed pollutants to the river, and the nutrient concentrations are high. The steelhead trout population is threatened.

Sweetwater Watershed

The watershed covers approximately 600 square kilometers and includes the Sweetwater River, Sweetwater Reservoir, and Loveland Reservoir. The upper regions contain parts of the Cleveland National Forest and Cuyamaca Rancho State Park. Multiple habitats (e.g., oak and pine woodlands, riparian forest, chaparral, coastal sage scrub, and coastal salt marshes). This is the most important watershed for water supplies and sensitive wetland habitats. The Sweetwater River flows into San Diego Bay at Sweetwater River Estuary. The estuary is tidal. The land is about 40% undeveloped. Most of the rest is split between residential and parks and recreation. Nearly 400,000 people live in the watershed. The major issues are surface and groundwater quality degradation, habitat degradation and loss, and invasive species. Coliform bacteria, trace metals, and other toxics result from agricultural and urban runoff.

Pueblo Watershed

This watershed has a great deal of residential, commercial, and industrial development. With the associated hard surfaces (e.g., concrete, asphalt), water running off into the creek contains nutrients, bacteria, and trash. Unfortunately, there is little remediation before the water reaches the Bay.

This watershed is the smallest (36,000 acres) and most urbanized in the region. More than 550,000 people live in it. It includes Chollas Creek and Paleta Creek and San Diego Bay. With those characteristics, it is no surprise that the major problem is urban runoff. The drainage includes relatively small creeks and pipes. Many are lined with concrete and drain directly into the Bay. Chollas Creek has some pollution with copper, lead, zinc, cadmium, and coliform bacteria, and some of those pollutants are toxic. Schiff, Bay, and Diehl (2003) examined the stormwater in the Bay. The plume covers about 2.25 square kilometers, and half of the plume is toxic to marine life. The authors examined the effects on reproduction in the purple sea urchin (*Strongylocentrotus purpuratus*). The toxicity was greatest near the mouth of the creek and decreased with distance upstream. Zinc was found to be the most toxic component.

Tijuana Watershed

The Tijuana River is extremely polluted. The treatment infrastructure is simply inadequate to the challenge. Gersberg et al. (2004) characterized the quality of the water during wet and dry seasons. They used a test for toxicity that involved a water flea (*Ceriodaphnia dubia*). Under baseline conditions, the toxicity was low, but it increased after periods of rain. Their tests suggest that nonpolar organics are the source of the toxicity. They also speculated that the first hard rain washed out the toxicity into the river so that subsequent rain events produced less toxicity.

Svejkovsky et al. (2010) used aerial imagery and other data to track stormwater runoff from the Tijuana River into the Pacific Ocean in 2003–2008. They could follow the runoff by its color on the surface of the ocean water. The discharge rate was the main determinant of fresh plumes. Bacterial levels were high in the fresh plumes. Oldest plumes contained both polluted and clean water. They concluded that high-resolution imagery is a valuable tool, but it cannot by itself identify polluted water plumes.

Ayad et al. (2020) studied the discharge of stormwater and wastewater from the Tijuana River into the Pacific. They used remote imaging to differentiate ocean plumes of stormwater, wastewater, open ocean, and mixes of those. They showed that stormwater had the highest levels of dissolved organic matter, turbidity (12.4 to 45.7 FNU), and enterococcus bacteria. Microplastics are another serious pollutant that reaches the water through runoff. De Jesus Piñon-Colin et al. (2020) looked at microplastics at seven sites on the Tijuana River. The largest number were found near an industrial site, and they were mostly polymers. Rainfall increased the numbers.

Filling and Dredging

While the Bay has remained roughly the same for millions of years, its depth and shape have changed considerably with filling and dredging (Canada, 2006). In the mid-19th century, the citizens of San Diego began dredging the Bay to open a channel for shipping. The course of the San Diego River was also changed into Mission Bay. Dredged material was used to create the Harbor and Shelter Island, and Coronado was restructured by filling Spanish Bight and Whaler's Bight.

Most of the wetlands around the Bay have been filled, diked, or drained. In the north and central bay, 90–100% of submerged lands, intertidal mudflats, and salt marshes have been filled and lost. In the South Bay, the San Diego Bay National Wildlife Refuge contains the last remaining large area of wetlands, mudflats, and eelgrass beds.

Mission Bay is one of the most manipulated areas of the San Diego Region. In 1923, Bonita Bay was dredged, and dredging and filing were used to build a causeway through the southern part of the bay and across the mudflats. In 1929, material was dredged from the bay to build a causeway across the bay from the north to Crown Point. The dredging operation improved boating on the bay. The dredging removed enough material to provide a new channel and form an oval course around the western bay. In 1946, dredging and filling began near Bonita Bay. Later dredging and filling projects further modified the Bay.

Mission Bay is not the only product of extensive dredging and filling. Originally, North Island and Coronado Island were separated by a shallow bay called the Spanish Bight. In 1945, the Bight was filled with dredged material, and the North

Island was connected to Coronado Island. Actually, Coronado Island was not an island, but a bit of land at the end of the peninsula called the Silver Strand.

In 2020, dredging was begun to expand a channel 5.5 km southeast of the Coronado Bay Bridge and near National City Marine Terminal and Sweetwater Channel. This is the first time the channel has been dredged since 1976. The project will remove 240,000 cubic yards of sediment from the channel. That material will be disposed of at several locations, including the LA-5 Ocean Dredge Material Disposal Site (175,000 cubic yards) and a site near Silver Strand State Beach (65,000 cubic yards). Dredging will also occur in northern San Diego, and that material will be used to build-up beaches in the city.

Sea Level Rise

Sea levels have varied greatly in the geologic past. In the late Pliocene (2.7 million years ago), much of the San Diego Area was underwater. Coronado, North Island and much of Point Loma were submerged (San Diego, 2019). At that time, the sea level was more than 60 meters higher than it is today. San Diego and Mission Bays are remnants of that time. During the Ice Age (about 15,000 years ago), near 30% of the Earth was covered with ice, and sea levels were much lower than the present. The shoreline was 12–35 km further west than it is today. San Diego Bay was dry.

About 10,000 years ago, the last Ice Age ended, and since then, the melting ice has been increasing sea levels. That rise was slow until the last century. Sea levels rose an average of 1.96 mm/year for a total increase of 20 cm (Curry, 2018). However, the world is now in the midst of another global warming scenario, and sea levels are beginning to rise again.

The San Diego Area is quite vulnerable to sea level rise. At the current rate of rise, much of the area will be flooded within the next 80 years. The worst-case scenario is apocalyptic. Much of the area's infrastructure would be lost, including 85% of wastewater pump stations, 60% of the city's marshland, 70% of the beaches, and 60% of hotels. Unfortunately, the situation is not promising. In the last century, sea level rise has been 2.5 cm per decade.

A better understanding of how sea level rise will affect coastal cities, such as San Diego and surrounding region are critical for planning. The effects of sea level rise will be compounded by the added effects of other hazards, such as heavy storms, coastal flooding, and high tides. It is important to look at all of these events together. Rahimi et al. (2020) did just that. Their results from Oakland, California, can be used in other areas. The effects of rising sea levels could also be modeled by the 2015–16 El Nino. The high sea levels and intense waves, along with low levels of rain, brought more sea water into estuaries than normal. Harvey et al. (2020) measured water in 13 estuaries in Southern California in that period. They found that a number of normally open estuaries formed a sand sill at the entrance of the estuary.

Inundation from sea level rise is complicated by other factors, such as subsidence and uplift (Dangendorf et al., 2017). These are caused by movement on geologic faults, mantle flow, and other processes. In addition, consolidation of sediments, biological processes, and human activities can cause compaction or uplift. Since humans have lived in the San Diego area, they have transformed the land around the Bay by filling low-lying areas. Those areas are still unstable and susceptible to subsidence. In addition, much of the marshland has been filled and developed. Thus, that buffer that provided a safety buffer against flooding is gone.

Tsunamis

Landslides and earthquakes do not occur only on land. They can happen in the Pacific Ocean, and when they do, they can set off a tsunami. Underwater landslides can be related to slides, slumps, reef failures, and earthquakes. Not all produce a tsunami, but some do. This is not just a theoretical possibility. Since 1806, more than two dozen tsunamis have hit San Diego. They result from both local and distant events. Many faults run parallel to the southern California coast and could produce a tsunami (Anderson et al., 1989). Areas near Del Mar and Palos Verdes are particularly prone to underwater landslides. Large landslides are estimated to occur about every 10,000 years. While distant earthquakes and landslides can also cause tsunamis in San Diego, these local events would occur with very little warning, making them even more dangerous.

Barberopoulou et al. (2011) examined the tsunami risks in San Diego Bay with realistic fault and landslide data. Fortunately, North Island and the Silver Strand protect the Bay from the worst of most tsunamis. However, tsunamis induce very strong currents, and those can do real damage inside the harbor. Most tsunamis generated at a distance result in waves of about 1 meter, but locally generated ones might overtop the Silver Strand. The inner harbor of the Bay is enclosed by the Point Loma Peninsula and by North Island, Coronado, and the Silver Strand. These geographic structures protect the inner harbor from large wave amplitudes but not tsunami currents. Over the last century, the harbor has been modified considerably. Space was made for mooring boars (private and Navy), but the reduced water surface area increased tsunami amplitudes and currents. Space was developed to moor both private and Navy boats. Unfortunately, the constricted channels reduced the water surface area and so increased tsunami amplitudes and currents. The channels were also dredged to make them deeper to improve navigation, but that also allows greater penetration of tsunami energy into the harbor. Those modifications had both positive and negative effects. They might constrain flooding outside the harbor and shunting high currents to areas that are of less value. Barberopoulou et al. (2015) studied these modifications and their effects. Their data will help planners to design harbors that are safer from tsunamis.

Watts (2004) reviewed the probabilities of such events off Southern California. The San Pedro Escarpment has suffered numerous failures (Bohannon and Gardner, 2004). The most recent was about 7500 years ago and involved the movement of 0.34 cubic kilometers of material. That landslide has all the characteristics of one that would have caused a significant tsunami of 3–12 meters.

BETWEEN LAND AND WATER

In the past, the wetlands were considered to be wastelands rife for development. Many were filled and their benefit was lost. In the last few decades, the importance of wetlands has been realized. Now there are efforts to save what wetlands remain and to reclaim areas that were once wetlands. Knowing where those wetlands existed in the past is very helpful in those efforts, and the best source of that information is the maps prepared by the U.S. Coast and Geodetic Survey many years ago. They are the best method for estimating the extent of the wetlands prior to the arrival of the Europeans.

Stein et al. (2014) have been examining these maps for Southern California. The early coastal estuarine habitats covered nearly 20,000 hectares. Moreover, 40% was salt marshes, 25% was salt or mudflats, and 35% was subtidal. Another 5000 hectares included dunes and beaches, wooded wetlands, high marshes, ponds, and river habitats. Over half of the estuarine habitats were in San Diego County, mostly around Mission and San Diego Bays.

Tidal and Mudflats

The floor of the Bay varies. At the mouth and along the western side, the floor is mostly sand. Deeper in the Bay and along the eastern side, the floor tends to be silt and clay.

The salt marshes in the San Diego area are accessible to fishes only about 16% of the time. West and Zedler (2000) surveyed five species in the Sweetwater Marsh National Wildlife Refuge for a year beginning in June 1997. *Fundulus parvipinis* and *Gillichthys mirabilis* were the most common species. The fish were well fed, demonstrating that the salt marsh provides excellent support for them. *F. parvipinis* was found to consume six times more food than fish species that remained in the adjoining creeks. These studies show the importance of restoring and maintaining salt marshes.

Methyl bromide and methyl chloride result from many sources, and salt marshes are one of them. Rhew et al. (2000) examined the vegetation at various heights in the salt marsh column at sites in the Mission Bay marsh and the San Dieguito lagoon. The marshes are enriched in salt from March to September. Tidal seawater floods them, and evaporation removes some of the water and leaves behind salt. In the wetter winters, rain dilutes the salt. The researchers found that the two compounds are produced by all cases, and in roughly the same ratios. Finally, although these habitats make up only a small portion of the Earth, by the scientist's calculations, they could account for nearly 10% of the total of these two compounds in the atmosphere.

Humans have destroyed most of the salt marshes. In Southern California, 75–90% of them are gone. Bromberg-Gedan et al. (2009) reviewed how centuries of human intervention has turned the salt marshes upside down. After all of that loss, one might wonder if those marshes could still sustain bird populations. Powell (2006) wondered just that. She looked at Belding's savannah sparrows (*Passerculus sandwichensis beldingi*) and light-footed clapper rails (*Rallus longirostris levipes*), two nonmigrating species that are sensitive to habitat. Loss of habitat and the arrival of new (i.e., red foxes or *Vulpes vulpes*) and old (i.e., common ravens or *Corvus corax*) predators also stress the birds. New habitat is not going to be made so it is not likely that the two species will survive.

Salt Ponds

At the southern end of San Diego Bay, salt has been produced since before 1872 (Figure 7.2). The operation involves allowing ocean water into the salt ponds and then closing them off. As the sun evaporates the water from the ponds, the concentration of salt increases. In the 1910s, they produced 36,000 metric tons of salt each year. Other companies produced bromine and magnesium, but they went out of business by 1950. Ownership of the salt ponds passed to the San Diego Regional Airport Authority and the US Fish and Wildlife Service, who leased it out for salt production, and that operation continues today, producing

FIGURE 7.2 Salt ponds of South San Diego Bay. Salt has been produced from seawater for many years. Seawater is sequestered in ponds, and the water is allowed to evaporate in the sun, leaving the salt behind (Photograph courtesy of the US Department of Agriculture).

about 68,000 metric tons per year. This and the salt operation in the South San Francisco Bay are the only such facilities in California.

The increased salt concentrations in the salt ponds are an ideal environment for brine flies (*Ephydridae* sp.) and brine shrimp (*Artemia* sp.). Those organisms cause a change in the color of the water to pale green and bright red. They also provide food for birds. Nearly 100 species of birds, including migratory birds, live in the salt ponds.

In recent years, projects have been undertaken to return some of the salt ponds to marshland. However, some of the organisms in the salt ponds are considered threatened and endangered, and that determination has limited the transformation.

AIR AND CLIMATE

Climate

San Diego has a Mediterranean climate with hot, sunny dry summers and cool, wet winters. The temperature varies from 50ºF to 77ºF. The average annual precipitation is. Snow is very rare and then only in the higher elevations inland. Fog is a common feature at the coast, especially in the summer months. Inland areas can be much warmer. The hottest temperatures often 30 cm accompany the Santa Ana winds, which blow from the east. These typically occur in the fall. The topography of the city with its hill and valley results in a number of microclimates.

As is the case with the rest of the world, the climate is changing in the San Diego Bay region. In fact, several changes are already felt, according to California's Fourth Climate Change Assessment (Kalansky et al., 2018). The report states that temperatures will increase by 5–10ºF by the end of the 21st century. Precipitation will be variable as it has been, but it will feature wetter winters and drier springs and more frequent and severe droughts. The risk of wildfires will increase. Sea levels will rise by 0.3 meters by 2050 and 1 meter or more by 2100. This will have serious consequences for inundating low-lying regions and threaten a lot of infrastructure. These changes will affect coastal properties, water supplies, and energy demands. They will also affect the area's plants and animals. For example, wetlands will be lost to the rising sea levels. Hotter temperatures and less water will also stress many animals. Summer fires will increase in severity. Those actions will be described in more detail below.

Atmospheric Rivers

Water has long been a serious concern in California. First, there is a serious imbalance in where water is found and where it is needed. About two-thirds of the rain and snow fall in the northern one-third of the state, but three-quarters of the population lives in the southern two-thirds of the state. Second, the state has a long history of alternating between drought and floods. Finally, as the state has grown in population, cities, farms, industry, and environmental concerns have competed for this scarce resource.

Atmospheric rivers are critical to California's water supply. They account for 20–50% of the annual rainfall, and thus, can mean the difference between drought and plenty of water in a state that needs more every year. Thus, understanding them is important for managing water resources in the state. In recent years, a number of studies have brought some clarity to this field. Atmospheric rivers are born in the eastern North Pacific where they result from low-level jets along the frontal edge of major winter cyclones. They are typically more than 2000 km long, a few hundred km wide, and in the lowest 2.5 km of the atmosphere. They stretch from California to Hawaii. Dettinger et al. (2011) developed models based on factors (e.g., sea-surface temperatures and atmospheric conditions) to predict atmospheric rivers. Guzman-Morales and Gershunov (2019) examined 16 climate models to assess their validity. While atmospheric rivers are most associated with the Northwest and even northern California, they also affect southern California. Harris and Carvalho (2018) examined atmospheric rivers that struck the West Coast between 1979 and 2013 and found that those in southern California differed in several key characteristics from those farther north.

Atmospheric rivers can be a blessing or a curse. Many provide badly needed rainfall to break droughts that occur often in California. Others result in excessive rainfall, damaging winds, and floods (Young et al., 2017).

Bay Air

For millions of years, the air above the San Diego Bay was pretty much the same. Volcanos periodically erupted to send large amounts of smoke and noxious gases into the air. Wildfires added to the pollution. Beyond that, change was rare. The arrival of humans changed that. The Native Americans and even the early Spanish colonists burned fires to prepare food, provide heat, and clear land, but their activities were limited. However, in the mid-19th Century, large numbers of Europeans arrived and air quality declined rapidly.

California is an important contributor to greenhouse gases in the United States. About 7% of total US greenhouse gases come from California. Transportation is the largest single contributor. The state has used legislation to attempt to reduce levels of these pollutants in 2050 to less than their levels in 1990. Those efforts have been effective in reducing levels of greenhouse gases and also in improving public health.

Air quality is a serious concern. Poor air quality is associated with poor health, increased deaths, and loss of productivity and quality of life. For example, the California Air Resources Board has estimated that a reduction of PM2.5 (emissions from burning gasoline, diesel, and wood), the state would avoid 7,200 premature deaths, 1,900 hospitalizations, and 5,200 emergency room visits each year (CARB, 2020). Air pollution has a dramatic effect on human health, particularly lung and heart disease (Brunekreef and Holgate, 2002).

In terms of air quality, San Diego and Tijuana are bound together by topography and meteorology (Quintana et al., 2015). One of the largest sources of air pollution are the maquiladoras (Blackman et al., 2004). These are assembly plants that produce goods for export, mostly to the United States. Of the 3000 maquiladoras in Mexico, 700 are in Tijuana. In addition, a number of power plants on the Mexican side of the border that supply electricity to San Diego contribute to air pollution (Bolorinos et al., 2018). Unfortunately, environmental laws are not as strict in Mexico, and American companies have been quick to capitalize on that fact.

San Diego is a busy port, and port activities can greatly influence air quality in the City. However, San Diego is also affected by the massive Los Angeles-Long Beach port region (Ault et al., 2009). Each year, ships in Los Angeles release 1.2–1.6 million metric tons of particulate material into the air. A significant portion of that pollution makes its way to San Diego. The material includes particles from soot, metals, sulfate, and nitrate, which are typical of the burning of residual oil from ships, refineries, and traffic.

Wildfires

In recent years, wildfires throughout California have made national news. Fires can be characterized as dominated fuel or wind, and they are influenced by geographical distribution, fire history, source of ignition, seasonal timing, resources at risk, and responses. Historically, wildfires in California have reflected the climate. In warmer, drier times, the number of fires was greater than in wetter, cooler times. This has been the case from 500 to 1800 CE (Marlon et al., 2012). Until large numbers of humans arrived, California was much more forested than it is now. Consequently, in some years, wildfires were quite large. Humans deforested much of the state, and the number of wildfires were reduced (Stephens et al., 2007). Over the past several decades, the number and size of wildfires in California and the San Diego area have increased. There are many reasons, including climate change, earlier springs, and longer droughts (Westerling et al., 2006). In 2007, the fires in San Diego County were particularly bad. Zauscher et al. (2013) examined the aerosols that were released in those fires. They found that most of the aerosols (84%) came from the burning of biomass and that the concentrations were high enough to pose a health risk to people in the county. The composition of the particles changed as environmental conditions changes. For example, potassium chloride and organic nitrogen were detected early on in the fires, and ammonium nitrate and amines appeared after the relative humidity increased.

The wildfires in San Diego County in 2003 released a large amount of gaseous and particulate pollutants (Viswanathan et al., 2006) (Figure 7.3). During the fire, levels of fine particulate matter exceeded the levels recommend by the United States government. Also levels of carbon monoxide, hydrocarbons, and methane. Ozone was not produced because the sunlight was blocked by the smoke. The numbers of people who became ill with respiratory problems and other ailments were much higher. Given that the climate crisis will increase the risk for wildfires, government agencies need to plan for those possibilities.

FIGURE 7.3 Wildfires. Wildfires are already a serious problem in California. They will become more common and more intense as global temperatures rise. This picture from the weekend of October 25, 2003, shows several fires whipped by Santa Ana winds. (Image courtesy Jacques Descloitres, NASA GSFC).

Smoke from the wildfires in California spread over a considerable part of the state. In the summer and fall of 2008, fires ravaged Northern California, and the smoke drifted all the way down to Southern California and the San Diego region. Hawkins and Russell (2010) sampled particles in the air in La Jolla and determined their composition and source. Was it from ships in the harbor, diesel trucks on the highway, or fires in Northern California? When there were no wildfires, the particles were mostly from fossil fuel burning. However, when there were fires, 74% of the particles came from burning biomass.

Climate change has multiple effects, and one of those will affect the well-known Santa Ana winds that increase the fire danger each year. Guzman-Morales and Gershunov (2019) studied the history of the winds in the 21st century. They found that all of the climate change models predict that the intensity of the winds will slowly reduce as climate change continues. They also found that the season for the winds is becoming shorter. The peak remains in November to January, but the number of wind events in the early and late season are lower than they were in the past.

Climate change also affects the amount of cloud cover and the shade that it provides. Without the cloud cover, vegetation suffers additional water stress and that can increase the potential for wildfires. Williams et al. (2018) examined 10 years of data to produce a model of cloud cover in Southern California. They found that the amount of cloud cover has decreased significantly since the 1970s and that it might be a key factor in the increase in wildfires in the area recently.

California's climate is changing due to global warming. Recent years have been much drier on average, and the state has had significant periods of drought. Higher temperatures for longer periods have greatly increased the risk of wildland fires and extended the fire season for additional months. Cloern et al. (2011) predict that these trends will continue throughout the 21st Century.

CONCLUSIONS

San Diego Bay is an extraordinarily beautiful natural treasure that was created by powerful forces, including tectonic plate movement, water, fire, and air movement. Those forces took millions of years to do their work. Humans have had less time, but they have changed many aspects of the natural environment and in most cases for the worse. The Native Americans were, in general, better stewards of our natural world. Yet even they have been implicated in the die-off of large mammals, and they certainly used some amount of fire to clear the land. The Spanish settlers introduced a number of "alien species" to the Americas, including cattle, horses, and pigs. However, the Europeans or Americans have made the most changes. The land has been paved and developed. Rivers have been dammed and channeled. The Bay has become polluted. We may be entering a sixth mass extinction that is caused by humans, and this one might threaten even our own species. Fortunately, there is a growing realization of the importance of the environment. That is detailed in Chapter 9.

REFERENCES

Allen LG, Findlay AM, Phalen M (2002) Structure and standing stock of the fish assemblages of San Diego Bay, California from 1994 to 1999. *Bulletin-Southern California Academy of Sciences* 101: 49–85.

Anders R, Mendez GO, Futa K, Danskin WR (2014) A geochemical approach to determine sources and movement of saline groundwater in a coastal aquifer. *Ground Water* 52: 756–768.

Anderson JG, Rockwell TK, Agnew DC (1989) Past and possible future earthquakes of significance to the San Diego region. *Earthquake Spectra* 5: 289–335.

Ault AP, Moore MJ, Furutani H, Prather KA (2009) Impact of emissions from the Los Angeles port region on San Diego air quality during regional transport events. *Environmental Science & Technology* 43: 3500–3506.

Auta HS, Emenike CU, Fauziah SH (2017) Distribution and importance of microplastics in the marine environment: A review of the sources, fate, effects, and potential solutions. *Environment International* 102: 165–176.

Ayad M, Li J, Holt B, Lee C (2020) Analysis and classification of stormwater and wastewater runoff from the Tijuana River using remote sensing imagery. *Frontiers in Environmental Science* 8: 240.

Barberopoulou A, Legg MR, Gica E (2015) Time evolution of man-made harbor modifications in San Diego: Effects on tsunamis. *Journal of Marine Science and Engineering* 3: 1383–1403.

Barberopoulou A, Legg MR, Uslu B, Synolakis CE (2011) Reassessing the tsunami risk in major ports and harbors of California I: San Diego. *Natural Hazards* 58: 479–496.

Barraza AD, Komoroske LM, Allen C, Eguchi T, Gossett R, Holland Daniel E, Lawson D, LeRoux RA, Long A, Seminoff JA, Lowe CG (2019) Trace metals in green sea turtles (Chelonia mydas) inhabiting two southern California coastal estuaries. *Chemosphere* 223: 342–350.

Bay SM, Greenstein DJ, Parks AN, Zeeman CQT (2016) Assessment of bioaccumulation in San Diego Bay. Southern California Coastal Water Research Project, SCCWRP Technical Report 953.

Benumof BT, Storlazzi CD, Seymour RJ, Griggs GB (2000) The relationship between incident wave energy and seacliff erosion rates: San Diego County, California. *Journal of Coastal Research* 16: 1162–1178.

Blackman A, Batz M, Evans D (2004) Maquiladoras, air pollution, and human health in Ciudad Juárez and El Paso. Discussion Papers dp-03-18. Resources for the Future. http://www.rff.org/documents/RFF-DP-03-18.pdf.

Blake AC, Chadwick DB, Zirino A, Rivera-Duarte I (2004) Spatial and temporal variations in copper speciation in San Diego Bay. *Estuaries* 27: 437–447.

Bohannon RG, Gardner JV (2004) Submarine landslides of San Pedro Escarpment, southwest of Long Beach, California. *Marine Geology* 203: 261–268.

Bolorinos J, Ajami NK, Muñoz Melendez G, Jackson RB (2018) Evaluating environmental governance along cross-border electricity supply chains with policy-informed life cycle assessment: The California–Mexico energy exchange. *Environmental Science & Technology* 52: 5048–5061.

Brandt JT, Sneed M, Danskin WR (2020) Detection and measurement of land subsidence and uplift using interferometric synthetic aperture radar, San Diego, California, USA, 2016–2018. *Proceedings of the International Association of Hydrological Sciences* 382: 45–49.

Brennecke D, Duarte B, Paiva F, Caçador I, Canning-Clode J (2016) Microplastics as vector for heavy metal contamination from the marine environment. *Estuarine, Coastal and Shelf Science* 178: 189–195.

Bromberg-Gedan K, Silliman BR, Bertness MD (2009) Centuries of human-driven change in salt marsh ecosystems. *Annual Review of Marine Science* 1: 117–141.

Brunekreef B, Holgate ST (2002) Air pollution and health. *The Lancet* 360: 1233–1242.

Canada L (2006) "Sitting on the Dock of the Bay" 100 years of Photographs from the San Diego Historical Society. *The Journal of San Diego History* 52: 1–17. https://sandiegohistory.org/journal/v52-1/pdf/2006-1_sitting.pdf

CARB (2020) Health and air pollution. California Air Resources Board. Retrieved on December 31, 2020, from https://ww2.arb.ca.gov/resources/health-air-pollution#:~:text=A%20number%20of%20air%20pollutants,some%20part%20of%20the%20year.

Carson RT, Damon M, Gonzalez JA, Johnson LT (2009) Conceptual issues in designing a policy to phase out metal-based antifouling paints on recreational boats. *Journal of Environmental Management* 90: 2460–2468.

Chadwick B, Leather J, Richter K, Apitz S, Lapota D, et al. (1999) Sediment Quality – Characterization Naval Station San Diego: Final Summary Report. Technical Report 1777. SSC San Diego. U.S. Navy.

Cloern JE, Knowles N, Brown LR, Cayan D, Dettinger MD, Morgan TL, et al. (2011) Projected evolution of California's San Francisco Bay-Delta-River system in a century of climate change. *PLoS One* 6(9): e24465.

Cohen-Waeber J, Sitar N, Bürgmann R (2013) GPS instrumentation and remote sensing study of slow moving landslides in the eastern San Francisco Bay hills, California, USA. *Proceedings of the 18th International Conference on Soil Mechanics and Geotechnical Engineering*, Paris. pp. 2169–2172.

Cordeira JM, Stock J, Dettinger MD, Young AM, Kalansky JF, Ralph FM (2019) A 142-Year climatology of Northern California landslides and atmospheric rivers. *Bulletin of the American Meteorological Society* 100: 1499–1509.

Curry J (2018) Sea level and climate change. Climate Forecast Applications Network Retrieved from: https://www.junkscience.com/wp-content/uploads/2019/02/special-report-sea-level-rise3.pdf

Dangendorf S, Marcos M, Woppelman G, Conrad CP, Frederikse T, Riva R (2017) Reassessment of 20th century global sea level rise. *Proceedings of the National Academy of Sciences USA* 114: 5946–5951.

De Falso F, Di Pace E, Cocca M, Avella M (2019) The contribution of washing processes of synthetic clothes to microplastic pollution. *Scientific Reports* 9: 6633.

Deheyn DD, Latz MI (2006) Bioavailability of metals along a contamination gradient in San Diego Bay (California, USA). *Chemosphere* 63: 818–834.

de Jesus Piñon-Colin T, Rodriguez-Jimenez R, Rogel-Hernandez E, Alvarez-Andrade A, Toyohiko Wakida F (2020) Microplastics in stormwater runoff in a semiarid region, Tijuana, Mexico. *Science of the Total Environment* 704: 135411.

Delgadillo-Hinojosa F, Zirino A, Holm-Hansen O, Hernández-Ayón JM, Boyd TJ, Chadwick B, Rivera-Duarte I (2008) Dissolved nutrient balance and net ecosystem metabolism in a Mediterranean-climate coastal lagoon: San Diego Bay. *Estuarine, Coastal and Shelf Science* 76: 594e607.

Dettinger MD, Ralph FM, Das T, Neiman PJ, Cayan DR (2011) Atmospheric rivers, floods and the water resources of California. *Water* 3: 445–478 10.3390/w3020445.

DeWyze J (2000) Toxic waste dumped in Mission Bay 1952-59. San Diego Reader. Retrieved from: https://www.sandiegoreader.com/news/2000/jul/20/cover-something-stinks-mission-bay/; accessed February 17, 2021.

EERI (2020) San Diego Earthquake Planning Scenario: Magnitude 6.9 on the Rose Canyon Fault Zone. Earthquake Engineering Research Institute, Oakland, CA, available at: https://sandiego.eeri.org/; accessed January 4, 2021.

Equinox (nd) Waste. Equinox Project. University of San Diego. Retrieved from: https://www.sandiego.edu/soles/hub-nonprofit/initiatives/dashboard/waste.php; accessed February 5, 2021.

Fairey R, Bretz C, Lamerdin S, Hunt J, Anderson B, Tudor S, Wilson CJ, LaCaro F, Stephenson M, Puckett M, Long ER (1996) Chemistry, toxicity, and benthic community conditions in sediments of the San Diego Bay Region Final Report. California State Water Resources Control Board. https://www.waterboards.ca.gov/water_issues/programs/bptcp/docs/reg9report.pdf; accessed January 9, 2021.

Field EH, and 2014 Working Group on California Earthquake Probabilities (2015) *UCERF3: A New Earthquake Forecast for California's Complex Fault System.* U.S. Geological Survey, Reston, VA.

Finnegan NJ, Broudy KN, Nereson AL, Roering JJ, Handwerger AL, Gennett G (2019) River channel width controls blocking by slow-moving landslides in California's Franciscan mélange. *Earth Surface Dynamics* 7: 879–894.

Garrick D (2019) San Diego plans to expand city dump despite zero-waste policy. *San Diego Union-Tribune.* Retrieved from: https://www.sandiegouniontribune.com/communities/san-diego/story/2019-10-25/san-diego-plans-to-expand-city-dump-despite-zero-waste-policy; accessed February 5, 2021.

Gersberg RM, Daft D, Yorkey D (2004) Temporal pattern of toxicity in runoff from the Tijuana River Watershed. *Water Research* 38: 559–568.

Gormlic F (2015) Why SeaWorld can't build a hotel at its location on Mission Bay. San Diego Free Press. Retrieved from: https://sandiegofreepress.org/2015/11/why-seaworld-cant-build-a-hotel-at-its-location-on-mission-bay/#.YC3DVGhKjIU; accessed February 17, 2021.

Greer K, Stow D (2003) Vegetation type conversion in Los Peñasquitos Lagoon, California: An examination of the role of watershed urbanization. *Environmental Management* 31: 489–503.

Guzman-Morales J, Gershunov A (2019) Climate change suppresses Santa Ana winds of Southern California and sharpens their seasonality. *Geophysical Research Letters* 46: 2772–2780.

Harris SM, Carvalho LMV (2018) Characteristics of southern California atmospheric rivers. *Theoretical and Applied Climatology* 132: 965–981.

Harvey ME, Giddings SN, Stein ED, Crooks JA, Whitcraft C, Gallien T, Largier JL, Tiefenthaler L, Meltzer H, Pawlak G, Thorne K, Johnston K, Ambrose R, Schroeter SC, Page HM, Elwany H (2020) Effects of elevated sea levels and waves on Southern California estuaries during the 2015–2016 El Niño. *Estuaries and Coasts* 43: 256–271.

Hawkins LN, Russell LM (2010) Oxidation of ketone groups in transported biomass burning aerosol from the 2008 Northern California Lightning Series fires. *Atmospheric Environment* 44: 4142–4154.

Hea L-M, He Z-L (2008) Water quality prediction of marine recreational beaches receiving watershed baseflow and stormwater runoff in southern California, USA. *Water Research* 42: 2563–2573.

Johnstone E, Raymond J, Olsen MJ, Driscoll N (2016) Morphological expressions of coastal cliff erosion processes in San Diego County. In: Brock JC, Gesch DB, Parrish CE, Rogers JN, Wright CW (eds.), *Advances in Topobathymetric Mapping, Models, and Applications. Journal of Coastal Research*, Special Issue, No.76, pp. 174–184. Coconut Creek (Florida), ISSN 0749-0208.

Kalansky, J, Cayan D, Barba K, Walsh L, Brouwer K, Boudreau D (2018) San Diego Summary Report. California's Fourth Climate Change Assessment. Publication number: SUM-CCCA4-2018-009.

Komoroske LM, Lewison RL, Seminoff JA, Deheyn DD, Dutton PH (2011) Pollutants and the health of green sea turtles resident to an urbanized estuary in San Diego, CA. *Chemosphere* 84: 544–552.

Komoroske LM, Lewison RL, Seminoff JA, Deustchman DD, Deheyn DD (2012) Trace metals in an urbanized estuarine sea turtle food web in San Diego Bay, CA. *Science of the Total Environment* 417–418: 108–116.

Lacroix P, Handwerger AL, Bièvre G (2020) Life and death of slow-moving landslides. *Nature Reviews Earth and Environment* 1: 404–419.

Lambert S, Wagner M (2016) Characterisation of nanoplastics during the degradation of polystyrene. *Chemosphere* 145: 265–268.

Latker A, Enright W, Gartman R (2020) Sediment. In: *Biennial Receiving Waters Monitoring and Assessment Report for the Point Loma and South Bay Ocean Outfalls 2018-19* (eds. R Kempster, AK Latker). City of San Diego. Chapter 4.

Masura J, Baker J, Foster G, Arthur C, Herring C (2015) Laboratory methods for the analysis of microplastics in the marine environment: recommendations for quantifying synthetic particles in waters and sediments. NOAA Technical Memorandum NOS-OR&R-48.

Marlon JR, Bartlein PJ, Gavin DG, Long CJ, Anderson RS, Briles CE, Brown KJ, Colombaroli D, Hallett DJ, Power MJ, Scharf EA, Walsh MK (2012) Long-term perspective on wildfires in the western USA. *Proceedings of the National Academy of Sciences USA* 109: E535–E543.

Martinez L (2006) Retired Mission Bay landfill investigated for hazards. SDNews.com. Retrieved from: http://www.sdnews.com/view/full_story/300536/article-Retired-Mission-Bay-landfill-investigated-for-hazards#:~:text=The%20site%20of%20the%20Mission,burying%20municipal%20and%20industrial%20wastes; accessed February 17, 2021.

Mason SA, Garneau D, Sutton R, Chu Y, Ehmann K, Barnes J, Fink P, Papazissimos D, Roger DL (2016) Microplastic pollution is widely detected in US municipal wastewater treatment plant effluent. *Environmental Pollution* 218: 1045–1054.

Mulkern AC (2019) Clifftop trains threatened by landslides in warming climate. *E&E News.* Retrieved from: https://www.eenews.net/stories/1061855765; accessed January 4, 2021.

Neira C, Cossaboon J, Mendoza G, Hoh E, Levin LA (2017) Occurrence and distribution of polycyclic aromatic hydrocarbons in surface sediments of San Diego Bay marinas. *Marine Pollution Bulletin* 114: 466–479

Neira C, Vales M, Mendoza G, Hoh E, Levin LA (2018) Polychlorinated biphenyls (PCBs) in recreational marina sediments of San Diego Bay, southern California. *Marine Pollution Bulletin* 126: 204–214.

Ponti DJ, Tinsley JC III, Treiman JA, Seligson H (2008) Ground Deformation, section 3c in Jones LM, Bernknopf R, Cox D, Goltz J, Hudnut K, Mileti D, Perry S, Ponti D, Porter K, Reichle M, Seligson H, Shoaf K, Treiman J, Wein A. The ShakeOut Scenario: U.S. Geological Survey Open-File Report 2008-1150 and California Geological Survey Preliminary Report 25 http://pubs.usgs.gov/of/2008/1150/.

Powell AN (2006) Are Southern California's fragmented saltmarshes capable of sustaining endemic bird populations? *Studies in Avian Biology* 32: 198–204.

Quintana PJ, Ganster P, Stigler Granados PE, Muñoz-Meléndez G, Quintero-Núñez M, Rodríguez-Ventura JG (2015) Risky borders: traffic pollution and health effects at US–Mexican ports of entry. *Journal of Borderlands Studies* 30: 287–307.

Rahimi R, Tavakol-Davani H, Graves C, Gomez A, Valipour MF (2020) Compound inundation impacts of coastal climate change: Sea-level rise, groundwater rise, and coastal precipitation. *Water* 12: 2776; DOI: 10.3390/w12102776.

Rhew RC, Miller BR, Weiss RF (2000) Natural methyl bromide and methyl chloride emissions from coastal salt marshes. *Nature* 403: 292–295.

Robbins EI, Quigley-Raymond S, Lai M, Fried J (2018) Microbial geochemistry reflecting sulfur, iron, manganese, and calcium sources in the San Diego River Watershed, Southern California USA. *Geosciences* 8: 495.

Rochman CM, Hentschel BT, Teh SJ (2014) Long-term sorption of metals is similar among plastic types: Implications for plastic debris in aquatic environments. *PLoS ONE* 9(1): e85433. https://doi.org/10.1371/journal.pone.0085433

San Diego (2019) City of San Diego State Lands Sea Level Rise Vulnerability Assessment. https://www.sandiego.gov/sites/default/files/state-lands-sea-level-rise-vulnerability-assessment.pdf; accessed January 9, 2021.

Schiff K, Bay S, Diehl D (2003) Stormwater toxicity in Chollas Creek and San Diego Bay. *Environmental Monitoring and Assessment* 81: 119–132.

SDState (nd) Learn about the San Diego Bay watersheds. College of Education. San Diego State University. Retrieved from: https://www.sdbay.sdsu.edu/education/watersheds.php; accessed February 5, 2021.

Stein ED, Cayce K, Salomon M, Bram DL, De Mello D, Grossinger R, Dark S (2014) Wetlands of the Southern California Coast – Historical extent and change over time. Southern California Coastal Water Research Project. SCCWRP Technical Report 826 SFEI Report 720.

Stephens SL, Martin RE, Clinton NE (2007) Prehistoric fire area and emissions from California's forests, woodlands, shrublands, and grasslands. *Forest Ecology and Management* 251: 205–216.

Svejkovsky J, Nezlin NP, Mustain NM, Kum JB (2010) Tracking stormwater discharge plumes and water quality of the Tijuana River with multispectral aerial imagery. *Estuarine Coastal and Shelf Science* 87: 387–398.

Thompson B, Melwani AR, Hunt JA (2009) Estimated sediment contaminant concentrations associated with biological impacts at San Diego Bay clean-up sites, SWRCB Agreement No. 08-194-190, Contribution No. 584, Aquatic Science Center, Oakland, California.

Turner-Tomaszewics C, Seminoff JA (2012) Turning off the heat: Impacts of power plant decommissioning on green turtle research in San Diego Bay. *Coastal Management* 40: 73–87.

Van A, Rochman CM, Flores RM Hill KL, Vargas E, Vargas SA, Hoh E (2012) Persistent organic pollutants in plastic marine debris found on beaches in San Diego, California. *Chemosphere* 86: 258–263.

Viswanathan S, Eria L, Diunugala N, Johnson J, McClean C (2006) An analysis of effects of San Diego wildfire on ambient air quality. *Journal of the Air & Waste Management Association* 56: 56–67.

Voosen P (2020) A muddy legacy. *Science* 369: 898–901.

Watts P (2004) Probabilistic predictions of landslide tsunamis off Southern California. *Marine Geology* 203: 281–301.

West JM, Zedler JB (2000) Marsh-creek connectivity: Fish use of a tidal salt marsh in Southern California. *Estuaries* 23: 699–710.

Westerling AL, Hidalgo HG, Cayan DR, Swetnam TW (2006) Warming and earlier spring increase western US forest wildfire activity. *Science* 313: 940–943.

White MD, Greer KA (2006) The effects of watershed urbanization on the stream hydrology and riparian vegetation of Los Penasquitos Creek, California. *Landscape and Urban Planning* 74: 125–138.

Williams AP, Gentine P, Moritz MA, Roberts DA, Abatzoglou JT (2018) Effect of reduced summer cloud shading on evaporative demand and wildfire in Coastal Southern California. *Geophysical Research Letters* 45: 5653–5662.

Wills CJ, Perez FG, Gutierrez CI (2011) Susceptibility to deep-seated landslides in California. California Geological Survey. https://www.conservation.ca.gov/cgs/Documents/Publications/Map-Sheets/MS_058.pdf; accessed January 4, 2021.

Wilson RC, Keefer DK (1985) Predicting areal limits of earthquake-induced landsliding. in JI Ziony, editor, Evaluating earthquake hazards in the Los Angeles region-an earth-science perspective: U.S. Geological Survey Professional Paper 1360, p. 317–345.

Wilson DS, McCrory PA, Stanley RG (2005) Implications of volcanism in coastal California for the Neogene deformation history of western North America. *Tectonics* 24: TC3008.

Wood M (2019) The city sends about 15% of the recycling it collects to the dump. *Voice of San Diego*. Retrieved from: https://www.voiceofsandiego.org/topics/news/the-city-sends-about-15-percent-of-the-recycling-it-collects-to-the-dump/; accessed February 5, 2021.

Young AM, Skelly KT, Cordeira JM (2017) High-impact hydrologic events and atmospheric rivers in California: An investigation using the NCEI Storm Events Database. *Geophysical Research Letters* 44: 3393–3401.

Young AP (2018) Decadal-scale coastal cliff retreat in southern and central California. *Geomorphology* 300: 164–175.

Young AP, Ashford SA (2006) Performance evaluation of seacliff erosion control methods. *Shore & Beach* 74: 16–24.

Young AP, Olsen MJ, Driscoll N, Flick RE, Gutierrez R, Guza RT, Johnstone E, Kuester F (2010) Comparison of airborne and terrestrial Lidar estimates of seacliff erosion in Southern California. *Photogrammetric Engineering & Remote Sensing* 76: 421–427.

Zauscher MD, Wang Y, Moore MJK, Gaston CJ, Prather KA (2013) Air quality impact and physicochemical aging of biomass burning aerosols during the 2007 San Diego wildfires. *Environmental Science & Technology* 47: 7633–7643.

Zeeman C, Taylor SK, Gibson J, Little A, Gorbics C (2008) Characterizing Exposure and Potential Impacts of Contaminants on Seabirds Nesting at South San Diego Bay Unit of the San Diego National Wildlife Refuge (Salt Works, San Diego Bay). FY05 Environmental Contaminants Program On-Refuge Investigations Sub-Activity. U.S. Fish and Wildlife Service Region 8.

8 Biology of the San Diego Bay Region

In this chapter, we describe the contemporary inhabitants of the San Diego region, in particular the trees, vegetation, marine, and estuarine plants and algae, as well as the major groups of animals, such as the keystone predators (for example, mountain lions and coyotes on land, and sea lions and whales in the sea) and their prey (deer, wild pigs, smaller predators, rodents and reptiles). The San Diego region is also a major region for overwintering and migrating birds and butterflies (the monarch, for example) and the Bay and estuaries of the San Diego watershed serve as sources of fresh water for the region's biota. Almost all the animal groups are discussed, including vertebrates and invertebrates, both terrestrial and aquatic. We also discuss the contemporary problems with the numerous invasive species that have been introduced, both deliberately and accidentally, by man over the past 200 years or so.

INTRODUCTION

The San Diego Bay region has a Mediterranean climate (CSa) to the north and a semi-arid climate (BSh) to the south and east, when using the Köppen climate classification system (Kottek et al., 2006). The climate is characterized in western San Diego by mild winters and warm dry summers, with most of the precipitation (more than 200 mm) occurring during the winter months (Federal Records, 2020; NOAA, 2011). The flora and fauna are typical of such a climate zone and many of the species endemic to the region are also to be found in similar climate zones in Eurasia (Jiang et al., 2019).

Regarding a description of the flora and fauna of the San Diego Bay region, it is not our intention to provide an encyclopedic list of all the resident species. Nor should this section be relied upon as a definitive accounting of the flora and fauna of the region. Rather, we will give a few chosen examples of how the environment has affected (and sometimes changed) the evolution of various organisms. As in Chapter 5, we will use the Linnaean binomial nomenclature where possible for additional study.

ANIMALS

Vertebrates

This section includes descriptions of all the extant animals that are classified as chordates, implying they have had at least a dorsal notochord and/or a nerve cord during ontogeny (embryonic growth and development) (Rychel et al., 2006).

DOI: 10.1201/9780429487460-8

As described earlier, the fauna of the Upper Pleistocene Epoch (129 to 11.65 kya) (kya: thousand years ago) were predominantly mammoths, mastodons, shrub ox, woolly rhinoceros, giant sloths, horses, camels, llamas, the monstrous short-faced bear, a number of large saber-toothed cats, American lion, American cheetah, and the dire wolf, in addition to the ancestors of those that survived into our present time, the Holocene Epoch (11.65 kya to the present).

We will introduce the description of the fauna of the San Diego region by way of class, size, and behavioral groupings. We will only include those animals which have in the past or currently occupy ecological niches in the region; we may also mention if a species that is considered endemic (a species confined to a specific ecological or geographical area) to San Diego is also found elsewhere, although this is very rare (for example salamanders of the genus *Ensatina*, which will be described in more detail later). We will describe in classic trophic terms, from the top predators through the herbivores and those smaller animals that depend upon both groups.

Mammals

Class Mammalia are descendants of the group of Mesozoic mammal-like reptiles, the therapsids, and which in fact significantly pre-date the better-known dinosaurs of the time (Kemp, 2005); they most likely evolved as incompletely endothermic reptiles from the pelycosaurs of the Permian Period (299 to 252 mya) (mya: million years ago) in response to a warming and dry climate (Kemp, 1987, 2005). As many readers will likely know, during the Cretaceous Period (145 to 66 mya) the ancestors of the extant mammals were usually quite small creatures, often living in burrows and being more active at night (Cifelli, 2000; Woodburne, 2004). Mammals are members of the clade Amniota, which are characterized by having a membrane, the amnion, surrounding the developing embryo. The ancestors of some present-day mammals in California were described in more detail in Chapter 5.

Predators

We first come to the Carnivora, the carnivores, which in turn comprise the following Families: Ursidae (bears), Canidae (dogs and their relatives), Mustelidae (weasels, martens, otters, badgers, skunks), Procyonidae (raccoons), Otariidae (eared seals), Phocidae (earless or true seals), and Felidae (all cats) (Wilson and Mittermeier, 2009; Flynn et al., 2005). The large or largest predators in an ecosystem are referred to by biologists and ecologists as a keystone predator. In this context, the term "keystone" refers to the influential niche position of the predator within the ecosystem and which contributes to how the ecosystem is maintained in balance (Beschta and Ripple, 2009; Wallach et al., 2010; Ripple et al., 2014). This is referred to as a top-down trophic cascade (Leopold, 1949; Hairston et al., 1960; Oksanen et al., 1981). Briefly, a keystone predator maintains an ecosystem in balance by predation upon herbivores that themselves feed upon vegetation critical to maintaining the structure of the ecosystem; for example, by securing the banks of a stream or river so that flooding and sedimentary buildup is controlled. If the keystone predator is removed, such as by human control measures or by disease, the prey herbivores multiply uncontrollably, devour more vegetation, and with less vegetation to secure the banks of the stream or river, the fluvial flow is compromised, leading to flow rate changes, build-up of silt, and eventual change of the topography and geomorphology of the affected area (Beschta and Ripple, 2009).

Although the large keystone predators (grizzly bear, *Ursus arctos californicus*) and grey wolves, (*Canis lupus*) had been hunted to extinction by the early 1900s (Miller and Waits, 2006) there still remain a number of keystone predators in the region. At one time, keystone predators were considered to be detrimental to livestock and danger to people, and so during most of the 18th, 19th, and 20th centuries they were allowed to be hunted at will, frequently with bounties for each claimed corpse; in California, hunting of mountain lions was not banned until 1990 under California Proposition 117 (California Fish and Game Code Div. 3, Ch. 9, Art. 2, 2785–2799.6). During the late 1980s and onwards, ecologists and other biologists determined that removing these keystone predators from the ecosystem led to significant damage to the environment (Edvenson, 1994; Beschta and Ripple, 2009).

Scientists now consider that the grizzly bear is really just a larger brown bear (also *Ursus arctos*) that migrated from Eurasia after the last Ice Age about 14 kya (thousand years ago) as they have no significant mitochondrial DNA differences (Cronin et al., 1991).

The remaining and extant large predators in California include the American black bear (*Ursus americanus californiensis* in the Sierra Nevada and *U. a. altifrontalis* in the north coast and Cascades), mountain lion (*Puma concolor*), also known as the puma, cougar, or catamount (probably an Anglicization of the Spanish for mountain lion, "gato monte"), and the coyote (*Canis latrans*) but, as mentioned above, brown bears are no longer found in the San Diego region. Black bears were never native to San Diego County; however, in 1933 twenty-eight black bears were introduced to the San Bernardino Mountains north of San Diego County and 11 to the San Gabriel Mountains. These now appear to have expanded their range to eastern San Diego County but so far are not close to the larger population centers on the coasts (Raftery, 2013).

In the San Diego Bay region, both cougars and coyotes have become more tolerant to humans, and encounters with coyotes within municipal boundaries and mountain lions in more rural areas are often reported in the local news (DMT, 2012). Residents at the periphery of the urban and suburban communities may often hear the squealing and yipping of coyote pups greeting their parents return after a successful night's hunt during the mid-summer.

One of the smaller common predators in the San Diego region is the gray fox (*Urocyon cinereoargenteus*) and which have become habituated to both urban and suburban life but are found throughout San Diego County (Ordeñana et al., 2010). They are considered to represent the most basal canids, that is, they are closest in morphology to the ancestor of all foxes, coyotes, dogs, and wolves

(Wayne et al., 1997). It is the only canid in the Americas that can climb trees (Fedriani et al., 2000; Sillero-Zubiri, Hoffman, and MacDonald, 2004).

Another more-rarely spotted predator is the red fox (*Vulpes vulpes* or *Vulpes vulpes fulva*) and these too have become habituated to urban and suburban life. Interestingly, these red foxes are not part of the fox clades inhabiting the southern (or montane) refugium (subalpine parklands and alpine meadows of the Rocky Mountains, the Cascade Range, and Sierra Nevada), but are descended from the North American eastern red fox, introduced to the lower part of California during the 1870s for the fur trade and fox hunting. It is considered to be more of a problem than the grey fox, as it indiscriminately hunts small birds, small rodents, and reptiles whereas the grey fox is more in harmony with the ecosystem it has evolved with, hunting rodent pests, lagomorphs, and insects; it is also frequently herbivorous (Fedriani et al., 2000).

The mountain lion (*Puma concolor*), also known as the puma or cougar, is to be found throughout the eastern foothills and there are also small populations found in the Torrey Pines State Natural Reserve, the Peñasquitos Canyon Preserve, Del Mar mesa, Escondido's Daley Ranch, Volcan Mountain Ecological Reserve, as well as the mountain ranges farther east (SDMMP, 2010; Brehme et al., 2014). Although frequently encountered in the San Diego region, there have been moves recently to temporarily protect pumas on the Central Coast through to Southern California under the Endangered Species Act (Sahagún, 2020). Their natural prey are deer, coyotes, galliform birds, such as turkey (*Meleagris gallopavo*) and the California quail (*Callipepla californica*), as well as rodents and insects (SDMMP, 2010).

The bobcat (*Lynx rufus*) is up to two times larger than the domestic cat, standing 38 cm (15"), and weighs about 9 kg (20 lb; females) and between 7 and 13.5 kg (16–30 lb; males) (CDFW, ND-a). They are present throughout the San Diego region but as they are generally nocturnal, are observed infrequently. During the day, to avoid larger predators (and humans) they rest in what is termed a "scrape", which is a shallow pit scraped out with their hind limbs. The bobcat is most likely descended from the Eurasian lynx, having migrated across the Bering Land Bridge (Beringia) about 2.6 mya (Johnson and O'Brien, 1997; Pecon-Slattery and O'Brien, 1998) but then became isolated during the subsequent Ice Ages.

The Musteloidea are a superfamily of the Order Carnivora and to which belong many of the small predators common in the San Diego region. They include the Procyonidae (raccoons and their allies, named for their original classification as "pre-dogs"), the Mustelidae (otters, badgers, weasels, and their allies, named for the musk (scent/must) gland used to mark territory), the Mephetidae (skunks), and the Ailuridae (the red panda).

The common raccoon *(Procyon lotor)* is the largest and the most prevalent member of the procyonid family in north America, but unlike most other carnivores, it is an opportunistic omnivore (Wozencraft, 2005). They are defined as a mesopredator, which are characterized by the rapid population growth of intermediate-sized predators in the absence of larger top predators (Martin, 2011).

The name "raccoon" is derived from the Algonquian word *aroughcoune*, meaning "he who scratches with his hands." The Algonquian nation (who call themselves *Omàmiwininiwak* and *Anicinàpe*) is in northeastern North America (present-day Québec and eastern Ontario) and who had traded raccoon furs with the French during the 16th to 18th centuries (Litalien, 2004; Poulter, 2010). Raccoons are closely related to the ringtail (*Bassariscus* sp.), found in the dryer parts of Oregon, Eastern California, and extending to the southwestern United States, the olingos (*Bassaricyon* sp.), and the coati (*Nasua* and *Nasuella* spp.) and are more distantly related to the kinkajou (*Potos flavus*), however, the latter species are only found in Central and South America (Wozencraft, 2005). It is largely nocturnal, digging up worms and small invertebrates, and small family groups are often seen wandering the suburbs of San Diego. Whilst they may hibernate during the winter in other parts of North America, the San Diego climate is sufficiently mild that raccoons active throughout the year. Although other procyonids are generally form social groups, raccoons differ in that, other than sow-kit groups, they tend to remain solitary and may engage in fights during encounters with other raccoons (Barrat, 2013).

Members of the family Mustelidae including badgers, otters (both sea and river subspecies), and weasels, are common throughout the San Diego region (Brehme et al., 2014; Lafferty and Tinker, 2014); these are also termed meso-predators and we will now describe them in more detail.

The American badger (*Taxidea taxus jeffersoni*), is not as common in much of the San Diego region, particularly in areas developed for human activity. They are, however, more abundant in the uplands and riverine localities in protected and private lands, such as in Rancho Guejito and the San Diego River/El Capitan Grande Reservation (WERC ND, Brehme et al., 2014). It is among the largest of the mustelids in North America, males reaching between 60 to 75 cm (23.5" and 29.5") in length and up to 8.6 kg (19 lb) in weight; the females, on the other hand, are generally smaller, at between 6.3 to 7.2 kg (14–16 lb) (Feldhamer et al., 2003). This is termed "sexual dimorphism" and is to be found throughout the mammalian class (Klymkowsky et al., 2016). It is an aggressive animal, although its young can be the target of the golden eagle (*Aquila chrysaetos*), mountain lion, coyote, or bobcat. It is primarily a fossorial carnivore (the word is derived from Latin, "fossa," a ditch), meaning that it hunts mammals and birds that usually reside in burrows or underground tunnel systems. Its prey in the San Diego region is the California ground squirrel (*Otospermophilus beecheyi*), the California vole (*Microtus californicus*), pocket gophers (*Thomomys bottae*), moles (*Scapanus latimanus*), and snakes, including rattlesnakes (*Crotalus* spp.).

The sea otter (*Enhydra lutris*) is native to the west coast of North America and is distinct from the Eurasian otter (*Lutra lutra*). It is the heaviest of the mustelids and unlike the others, has no scent gland (Kenyon 1969). Sea otters make their home amongst the giant kelp forests growing in the shallow waters of the coast (see above) and its shelter provides much of the food for them, mainly echinoderms, such as sea urchins, crustaceans, mollusks, and fish, and it is a keystone species (Grenfell, 1974; Melquist and Dronkert, 1987; Larsen, 1984).

Another common mustelids resident in the San Diego region is the long-tailed weasel (*Mustela frenata*) and which is generally found in the more rural parts of the San Diego region.

The striped skunk (*Mephitis mephitis)* was once included within the mustelids, but is now classified in its own family, Mephetidae, and like the raccoon, is an omnivore (Dragoo and Honeycutt, 1997). Like the raccoon, it is primarily nocturnal although it can be easily detected during the daylight hours by its pungent smell, originating in its anal scent glands, situated ventral to its tail.

A visitor to the beaches, Mission Bay, and the cliff-tops of San Diego is unlikely to miss the large marine predator, the California sea lion (*Zalophus californianus*). It is in the family Otariidae, the eared seals, having external ear flaps, and which differ from true seals, which lack external ears. They are also distinguished from the true seal in that their front flippers are long and strong enough to hold themselves upright and they have rear flippers which are used for propulsion on land, facing forwards.

Another rare visitor to the San Diego region is the northern fur seal (*Callorhinus ursinus*), also in the family Otariidae (eared seals), but these rarely come ashore on the mainland and are mainly found on San Clemente or out at sea, hunting. They are a protected species under the Marine Mammal Protection Act of 1972.

True seals, on the other hand, are represented on the San Diego coastline by the Pacific harbor seal (*Phoca vitulina*). They are more specialized for aquatic life than are sea lions, having lost the ability to ambulate using their flippers on land.

Another rare visitor to the San Diego region is the northern elephant seal (*Mirounga angustirostris*) and which has become a more common sight during the past 30 to 40 years in California than in the past (Abadía-Cardoso et al., 2017) when they became protected under the Marine Mammal Protection Act. Since then, their numbers have increased spectacularly and have probably reached the population size before they became targets for their oil.

Prey

The vast majority of large prey mammals familiar to residents of the San Diego region are deer, predominantly the Columbian black-tailed deer (*Odocoileus hemionus columbianus*) and which normally inhabit a 100-mile-wide band of woodlands and chaparral-covered coastal mountains extending inland from the Pacific Ocean. They are often mistakenly called "mule deer" but are a separate subspecies related to the California mule deer (*Odocoileus hemionus californicus*) found on the western flanks of the Sierra Nevada range and in the hills and mountains to the east of San Diego (for example, Cuyamaca Rancho State Park), so called for their large ears. The most southerly population range of the black-tailed deer overlaps with that of the mule deer and hybridization between the two species is extensive.

European wild boar (*Sus scrofa scrofa*) were introduced into Monterey, California, for hunting during the 1920s; its domesticated descendant, common swine or pigs were brought over by Spanish settlers during the 1700s and many have since become feral and the present-day population is a wild boar/swine hybrid (Woodward and Quinn, 2011; Mayer and Brisbin, 2008). Their populations are generally located in eastern San Diego County and are not common in the San Diego region.

Mid-sized and small prey mammals, usually rodents, are commonly seen in all parts of the San Diego region. Some examples are: the California ground squirrel (*Otospermophilus beecheyi*); the Western grey squirrel (*Sciurus griseus nigripes*); the fox squirrel (*Sciurus niger*), which appears to have been introduced to Southern California from the east coast around 1904 (King, 2004; Ortiz and Muchlinski, 2014), and which now is the predominant species in the San Diego region over the Western grey squirrel; native dusky-footed woodrats (*Neotoma fuscipes*); the brown rat (*Rattus norvegicus*, introduced 1750–1755; Norwak, 1999); the more common black rat (*Rattus rattus*), observed often in yards and may inhabit attic space (so-called “roof rats”), and which was probably introduced during the early- to mid-1800s whose mtDNA is more closely related to rats from South and South-east Asia (Lantz, 1909; Conroy et al., 2012); the California vole (*Microtus californicus*); mice (*Mus musculus domesticus*); the ornate shrew (*Sorex ornatus*); moles, including the American shrew mole (*Neurotrichus gibbsii*), the broad-footed mole (*Scapanus latimanus*), and the coast or Pacific mole (*Scapanus orarius*); rabbits, including the black-tailed jackrabbit (*Lepus californicus californicus*) and the brush rabbit (*Sylvilagus bachmani*).

Another important rodent that claims the San Diego region as home is the golden beaver, (*Castor canadensis subauratus*); it was introduced by the California Department of Fish and Game during the 1920s although it had been apparently common up until at least the mid-1800s, but became extirpated (locally extinct) probably due to overuse of the Sweetwater River (which drains into the San Diego Bay) for agricultural irrigation (Graves, 1964; Hensley, 1946).

One bane of San Diego region residents is the Botta’s pocket gopher (*Thomomys bottae*); in their natural habitat, they create a network of tunnel systems that provide protection and a means of collecting food. However, when they reside in peoples’ gardens, they can cause havoc and destroy lawns, small shrubs, and entire horticultural endeavors.

Whales

In the last century, the Pacific gray whale (*Eschrichtius robustus*) was hunted nearly to extinction. Between 1846 and 1874, as many as 8,000 gray whales were killed by whalers. Other hunts continued until 1936 when gray whales became protected by the US (Rice, 1998). By the 1940s, there was sufficient concern amongst the scientific and many western industrial communities that a moratorium on hunting should be enacted. The International Whaling Commission (IWC) enacted such a moratorium in 1946 and 1982, although some nations and aboriginal communities were permitted to continue at much reduced levels (“The Economist” article, 2012). Since those times, baleen whale populations have increased significantly and residents of the San Diego region are able to observe up close migrating grey whales (*Eschrichtius robustus*), minke whales (*Balaenoptera acutorostrata*), blue whales (*Balaenoptera musculus*), and sperm whales (*Physeter catodon*) at many times of the year. More rarely, humpback whales (*Megaptera novaeangliae*) are seen.

In addition to the magnificent immense baleen whales, smaller toothed whales, such as dolphins (Delphinidae) and porpoises (Phocoenidae) are also frequently observed coursing back and forth along the coast, preying upon schools of fish, usually sardines or anchovy.

Marsupials

The only marsupial found in North America is the Virginia opossum (*Didelphis virginiana*). This ancient pouched mammal lives for only about three years but breeds when it is six (for a female) or eight (male) months old up to three times a year. They are generally nocturnal but are a familiar sight even in urban environments, especially when food is left outside for pets; they are omnivorous (Krause and Krause, 2006). The tail is prehensile and young opossums use this to grip onto the mother's back. As described in Chapter 5, the opossum's presence in North America resulted from the Great American Biotic Interchange (GABI) just under 3 million years ago.

Reptiles and Their Allies

Class Reptilia include all the predominantly exothermic tetrapods that usually lay eggs on land, eggs that are protected by a gas-permeable, proteinaceous, and/or calciferous shell surrounding and protecting the amnion. They represent the group of tetrapods that successfully adapted and emigrated from the aquatic environment to the terrestrial environment during the Carboniferous Period, the amniotes (359 to 299 mya). Living members include lizards, snakes, crocodilians, and birds.

Birds

Clade Aves now is considered to be placed within the larger clade of Dinosauria, and they are generally referred to by taxonomists as avian dinosaurs (Ostrom, 1973; Padian, 1986).

Land Birds

For reasons of space, we cannot describe each bird species found in the San Diego region and we suggest the reader finds other more comprehensive resources, such as the many ornithological handbooks available to the general public. Here we describe some of the birds that are of particular significance to the ecology of the San Diego region.

The bird that has captured the most attention in the minds of Californians during the past few decades has been the re-introduction of the largest bird of prey in North America, and the largest of the North American New World vultures, the California condor (*Gymnogyps californianus*), second only in size to the Andean condor (*Vultur gryphus*).

FIGURE 8.1 California condor. The birds were photographed at the San Diego Zoo, which was a key part of the effort to bring the condor back from the brink of extinction. Since the program began, the population of condors has grown from 22 to more than 500. Photograph reproduced with permission from Dr. Cynthia Wikler.

Extinct in the wild since 1987, the then-captive population of some 22 individual birds were the subject of an intensive captive breeding program, the California Condor Recovery Plan, led by the San Diego Wild Animal Park and the Los Angeles Zoo (CDFW, ND-b, San Diego Zoo News report ND) (Figure 8.1). One innovative feature of the program was that no humans were allowed to be seen by any hatchlings, to not become habituated to humans; the condor chicks were fed small portions of meat by a puppet condor consisting of the head and neck, within which the scientist/keeper's hand and arm were hidden. Mixed-age California condors were released in 1991 and 1992 in California at Big Sur, Pinnacles National Park, and Bitter Creek National Wildlife Refuge (southwestern San Joaquin Valley), and the first breeding pairs were observed establishing a nest in 2006. This is due to the late age at which they begin to reproduce (6 years old) as well as the fact that condors will only raise a single chick every other year (FWS, 2007).

According to the latest information from the United States Fish and Wildlife Service, there are approximately 160 California condors present in Central and Southern California and to more than 500 in the United States (FWS, 2020).

However, recent wildfires in California may have contributed to loss of some of these magnificent birds, including at least one chick (Rubenstein, 2020).

The turkey vulture (*Cathartes aura*), sometimes called a turkey buzzard, is another member of the New World vultures and is the most common scavenging bird seen in the skies of the San Diego region; as would be expected, they are often seen soaring and circling up using updrafts high into the sky to better see over the vast plains and hills of southern California. They are also to be found in the middle of the highway, a small group tearing at the carcass of road-kill.

The San Diego region has a large number of native birds of prey, in particular since the environment provides habitat for so many of their typical prey. These include the great horned owl (*Bubo virginianus*); the barn owl (*Tyto alba*); the burrowing owl (*Athene cunicularia*), the western screech owl (*Megascops kennicottii*), the bald eagle (*Haliaeetus leucocephalus*) and golden eagle (*Aquila chrysaetos*), the osprey (*Pandion haliaetus*), the prairie falcon (*Falco mexicanus*), the American peregrine falcon (*Falco peregrinus anatum*), the red-tailed hawk (*Buteo jamaicensis*), northern harrier or marsh hawk (*Circus cyaneus*), and white-tailed kite (*Elanus leucurus*) (Robbins et al., 1983; Sibley 2000). The bald eagle was almost extinct due to heavy use of pesticide contamination of their prey, but are now more commonly seen, as they migrate to and from Canada, Montana, Wyoming, and Idaho during the spring and fall (FS ND). Global warming may also one day affect the habitat that the bald eagle enjoys in southern California; the Audubon Society has calculated that if global temperatures warm by 1.5 °C, much of the winter range of the birds in coastal California including San Diego Bay, may be lost (Audubon ND).

The crow family in the San Diego region is distinguished by the common raven (*Corvus corax*), the American crow (*Corvus brachyrhynchos*), and the California or western scrub-jay (*Aphelocoma californica*), and its subspecies, the coastal western scrub-jay (*Aphelocoma californica californica*), all of which are predominantly woodland and scrub birds (Robbins et al., 1983).

As mentioned in Chapter 4, the San Diego region is a critical stopover point for migratory birds along the Pacific Flyway migration route, in particular, the shallow waters of Mission Bay and the Tijuana River estuary, which are a welcome rest point for shorebirds and waterfowl, as well as for sparrows and thrushes on the land, and which can number of up to one million birds.

Some of the other native birds and waterfowl are listed here: the greater roadrunner (*Geococcyx californianus*), the American goldfinch (*Carduelis tristis*), the American robin (*Turdus migratorius*), Brewer's blackbird (*Euphagus cyanocephalus*), red-winged blackbird (*Agelaius phoeniceus*), yellow-headed blackbird (*Xanthocephalus xanthocephalus*), California towhee (*Melozone crissalis*), spotted towhee (*Pipilo fuscus*), dark-eyed junco (*Junco hyemalis*), great egret (*Casmerodius albus*), snowy egret (*Egretta thula*), mallard (*Anas platyrhynchos*), Canada goose (*Branta canadensis*), snow goose (*Chen caerulescens*), and hummingbirds, including Anna's hummingbird (*Calypte anna*), Costa's hummingbird (*Calypte costae*), and the rufous hummingbird (*Selasphorus rufus*) (Robbins et al., 1983). The European or common starling (*Sturnus vulgaris*) is an invasive species, but giant flocks are to be seen in the early evenings, where they gather before nesting prior to sunset. Examples of other non-native birds are the red-masked parakeet (*Aratinga erythrogenys*), the budgerigar (*Melopsittacus undulatus*), and the red-crowned parrot (*Amazona viridigenalis*) (Stacey ND).

The Californian turkey, *Meleagris californica*, is an extinct species of turkey indigenous to the Pleistocene and early Holocene of California. It became extinct about 10,000 years ago (Bochenski and Campbell, 2006). The present Californian wild turkey population derives from wild birds re-introduced from other areas by game officials, a previous introduction program dating from between the 1920s and early 1950s having been unsuccessful. The Rio Grande wild turkey (*M. gallopavo intermedia*) was introduced throughout California during the 1960s and 1970s, whereas the Eastern wild turkey (*M. g. silvestris*) was released along the northern coast and has since formed a hybrid species with the Rio Grande subspecies (California Department of Fish and Game, 2005).

Birds frequently seen around the bay are the brown pelican (*Pelecanus occidentalis*), its subspecies the California brown pelican (*Pelecanus occidentalis californicus*), the American white pelican (*Pelecanus erythrorhyncas*), the double-crested cormorant (*Phalacrocorax auritus*), and Brandt's cormorant (*Phalacrocorax penicillatus*), all of which feed on fish by dive-bombing into the water from a height of usually about 5–10 m. They are often observed fishing together but since they are adapted to prefer different fish at different depths, there is little competition for food.

For further research, the reader is recommended to use some of the many well-produced handbooks available in the general press.

Seabirds and Shorebirds

One ecosystem in the San Diego region that we have not yet discussed is that of the peri-marine environment, the shoreline and the areas adjacent to them. It is here that numerous gulls and other sea and shorebirds compete for food and nesting space. There are many recognized species within the gull group (Laridae) although genetic analysis often demonstrates hybridization between 'species' and that most have up to 98.7% DNA in common (Paton, 2003; Pons, Hassanin, and Crochet, 2005; van Tuinen, Waterhouse, and Dyke, 2004). The following lists the more common gulls seen in the San Diego region, not only at the seashore, but also farther inland, where they may be seen raiding trash heaps up to 200 km from the ocean (Ackerman and Peterson, 2017). Included in no particular order are the Western gull (*Larus occidentalis*), the California gull (*L. californicus*), the glaucous-winged gull (*L. hyperboreus*), the herring gull (*L. argentatus*), Thayer's gull (*L. glaucoides thayeri*), Heermann's gull (*L. heermanni*), and Boneparte's gull (*Chroicocephalus philadelphia*). Some eat brine shrimp, grasshoppers, seal afterbirths, fish, squid, carrion, small mammals, and the eggs and chicks of other seabirds – even those of their own kind.

Other smaller seabirds include the common tern (*Sterna hirundo*), the gull-billed tern (*Gelochelidon nilotica*), the California least tern (*Sternula antillarum browni*), and Sabine's tern (*Xema sabini*), all of which spend more time at sea compared with that of the gulls. On the other hand, waders or shorebirds spend all their feeding time scurrying along the flat beach just along the tidemark, darting here and there looking for worms, small crustaceans, and other arthropods. Examples of such waders are the western sandpiper (*Calidris mauri*), the American avocet (*Recurvirostra americana*), the black oystercatcher (*Haematopus bachmani*), and Wilson's plover (*Charadrius wilsonia*) (Audubon California, ND).

The Tijuana River Estuary is one of the largest undeveloped coastal wetlands and salt marshes in Southern California and is home to many native birds as well as being an important stopover point on the Pacific Flyway and has been designated a wildlife sanctuary, the Tijuana River National Estuarine Research Reserve (Carlisle et al., 2009; Hertzog, 1990). One noted resident is Ridgeway's rail (*Rallus obsoletus*), which is also found in the San Francisco Bay to southern Baja California; it is a near-threatened species (BirdLife International, 2016).

Snakes and Lizards

Snakes and lizards (clade Reptilia) are amniotes that lay their eggs on land, the eggs having a gaseous-permeable, proteinaceous shell; unlike birds, they are exothermic and must warm under sunlight before they are able to hunt or forage.

Snakes

Let us begin with some of the snakes (Infraorder Serpentes) that inhabit the San Diego region. The relatively warm and dry climate, offered a multiplicity of habitats for small birds, mammals, reptiles, and invertebrates, the common prey for most snakes.

Garter snakes are some of the more common small (< 1 m) snakes found almost everywhere in the San Diego region, each species specialized to a particular habitat (California Herps, 2020a). The California red-sided garter snake (*Thamnophis sirtalis infernalis*) is such an example. Their prey are generally red-legged frogs and juvenile bullfrogs. Interestingly, garter snakes are one of the few animals capable of ingesting the toxic California newt (*Taricha torosa*) without incurring sickness or death (California Herps, 2021; Williams and Brodie, 2003).

The two-striped garter snake (*Thamnophis hammondii*) is most aquatic of the garter snakes; it is found throughout the region where there are water sources. Its diet includes fish, fish eggs, tadpoles and newt larvae, small frogs and toads, and earthworms (California Herps, 2021).

The California striped racer (*Coluber lateralis lateralis*) is frequently mistaken for the aquatic garter snake in that its upper body is predominantly colored black and has two lateral yellow stripes along the length of the body. Its diet includes live animals such as insects, lizards, snakes, birds, and even small mammals (Stebbins, 2003; Jennings, 1983). It is endemic to California and is currently listed as endangered, since its habitat is being destroyed by industrial and urban development.

A common snake often found sunning itself on the tarmac of a quiet byway, is the Western yellow-bellied racer (*Coluber constrictor mormon*). They are predominantly found in the foothill grasslands, brushlands, and moist environments throughout the region. They eat mice, fledgling birds, and lizards and are non-venomous (Murray, 2004).

Another common snake is the San Diego gopher snake (*Pituophis cantenifer annectens*) and is found throughout the coastal and hinterland of southern California (California Herps, 2021).

The Southern Pacific rattlesnake (*Crotalus oreganus helleri*) is not uncommon in the San Diego region; it is to be found predominantly in the hills and undeveloped valleys that surround the residential suburban torus that surrounds the San Diego Bay as well as in the foothills to the east. It is extremely venomous and any bite should be treated as potentially life-threatening.

One snake that is rarely seen in the San Diego region is the California kingsnake (*Lampropeltis getula californiae*). They may be found under logs and rock outcrops throughout the eastern and foothill regions and their prey included other snakes, (including rattlesnakes) lizards, birds, eggs, and small rodents. They restrain and kill their prey as a constrictor, wrapping their body around the prey and suffocating it before ingesting the animal headfirst. They are immune to rattlesnake venom.

Lizards

There are at least 20 species of native lizards (Order Squamata) that inhabit all the different ecosystems of the San Diego region; there are also a few introduced species (California Herps, 2020b). Those native include the alligator lizards (*Elgaria* spp.), the San Diegan legless lizard (*Anniella stebbins*), the granite spiny lizard found predominantly at higher elevations (*Sceliporos orcutti*), and the San Diegan tiger whiptail (*Aspidoscelis tigris stejnegeri*). Introduced species include geckos (*Hemidactylus* spp. and *Tarentola mauritanica*) and anoles (*Anolis* spp.) (California Herps, 2021).

The most commonly seen lizards, usually sunning themselves on a rock in backyards, are the woodland alligator lizard (also called the San Diego alligator lizard; *Elgaria multicarinata webbii*), a subspecies of the southern alligator lizard (*Elgaria multicarinata*). The term "alligator" refers to the presence of supporting bony structures in their dorsal and ventral scales, as found in alligators. They can reach a size of over 30 cm including the tail (Stebbins, 2003).

Turtles, Terrapins, and Tortoises

Turtles

A turtle native to the San Diego region is the southwestern pond turtle (*Actinemys pallida*). It is an endangered species and is only known from a tributary of the Sweetwater River near Jamul, east of Rancho San Diego (Kucher, 2010).

The two other turtles that may be seen in rivers, streams, and ponds in the San Diego region are both invasive. One is the red-eared slider (*Trachemys scripta elegans*; alien) is native to ponds and shallow streams from New Mexico eastwards throughout the mid-west, much of the south, and Appalachia. The other is the Western painted turtle (*Chrysemis picta bellii*) whose original range were the rivers and waterways of the mid-west and the northern prairie. Those resident in southern California are most likely descendants of pets that had been abandoned by their owners (California Herps, 2021).

Amphibians

Class Amphibia include salamanders, newts, frogs, and toads, as well as a limbless amphibian group, the caecilians. The largest of the San Diego region amphibians is the western tiger salamander (*Ambystoma mavortium*) up to 25 cm in length, but it is an invasive species probably introduced from the Great Plains; it is found only in small pockets of the foothills east of Lakeside. Of particular interest, this is one of the many species of salamanders which can, under certain environmental conditions, retaining the larval form whist becoming sexually mature; this is termed "neoteny" and is considered in the tenets of biological sciences to be one way that evolution to adapt to changing conditions and speciation may occur, including in humans (Kucera, 1997; Gould, 1977). A neotenous salamander that may be familiar to the reader is the Mexican axolotl (*Ambystoma mexicanum*) and which has been frequently kept in aquariums as a pet; it frequently remains in the larval form throughout its life, which can be up to 25 years (California Herps, 2021).

A salamander frequently spotted in the damp woodlands in the hills and mountains of California and southern Oregon is a lungless salamander, the common ensatina, *Ensatina eschscholtzii*, and which has been the subject of continuous scientific study at the University of California at Berkeley since the 1940s. It was one of the first animals to be identified as a ring species, that is, the subspecies are distributed around a geographic feature and they eventually meet up at either end of the ring as two non-interbreeding species whereas those subspecies found in adjacent habitats "around the ring" could interbreed in what is termed a "hybrid zone" (Stebbins, 1959; Brown and Stebbins, 1964).

Briefly, Stebbins determined from his studies of coloration and morphology that the seven different groups of salamander which encircled the California Central Valley were most likely each a different subspecies of the ensatina salamander and that the two subspecies in the Tehachapi Mountains, *E. e. eschscholtzii* (Monterey ensatina; western subspecies) and *E. e. croceater* (yellow-blotched ensatina; eastern subspecies) could not interbreed. Further south, in the Laguna Mountains, Santa Rosa Mountains, San Jacinto Mountains, and the San Bernardino Mountain ranges, another non-interbreeding population is found: *E. e. klauberi*. These findings were later borne out by enzyme polymorphisms and genetic data (Moritz et al., 1992; Jackman and Wake, 1994; Wake, 1997). However, there are some reports of *E. e. eschscholtzii/klauberi* hybrids on Palomar Mountain; in most cases (~90%) the hybrids have *E. e. klauberi* mitochondrial DNA, suggesting that these relatively rare pairings were limited to female *E. e. klauberi* and male *E. e. eschscholtzii* (Devitt, Baird, and Moritz 2011). The predominant subspecies of the San Diego region is *E. e. eschscholtzii* (Stebbins, 1959; Wake, 1997).

With the exception of the Lassen Gap zone, in almost all the regions where two subspecies' ranges adjoin each other, there is a zone of hybridization, which may only extend over a band of 730 to 2000 meters across (Alexandrino et al., 2005; Kuchta et al., 2009; Pereir and Wake, 2009). This parameter conforms to the mean territorial range of the salamanders, about 19 m, in that the overlapping territories of neighboring salamanders would necessitate an intervening population numbering about 5–10 of each subspecies on either side of the hybridization zone (Stebbins, 1954).

An example of another lungless salamander is the garden slender salamander (*Batrachoseps major major*); it inhabits the coastal sage scrub and the hills surrounding San Diego and is notable from its long slender body (up to 12 cm including the tail) and its tiny limbs, disproportionately small for its size. Its coloration can vary from between very pale grey to sandy or ruddy grey. A related species, the desert slender salamander (*Batrachoseps major aridus*) inhabits more arid and mountainous regions east of San Diego but survives as it can take advantage of sudden flash floods that leave long-lasting pools. Although similar in size, it can be distinguished from the garden slender salamander by its darker, almost gun metal grey, coloration (California Herps, 2021).

The California newt (*Taricha torosa*) inhabits only a small area deep within the San Bernardino Mountains, east of San Diego. There is considerable evidence that the coloration pattern of the *Ensatina schscholtzii xanthoptica* is a mimic of the California newt (Kutcha et al., 2008; Tan and Wake, 1995).

The California red-legged frog (*Rana draytonii*), now an endangered species, was once a very common sight in the 19th century and was made famous in Mark Twain's *The Celebrated Jumping Frog of Calaveras County*. Its loss is most likely due to competitive pressure from the invasive American bullfrog, the latter introduced for its legs as a food source. Previously the red-legged frog had been the source of frogs-legs in the state (Stebbins and McGinis, 2012). It had not been seen in San Diego County since 1974, but has been reintroduced to the Wheatley Ranch (Mesa Grande, east of Escondido) using funds provided by the US Fish and Wildlife Service, The Nature Conservancy, San Diego Association of Governments, and the Wheatley family (Heil, 2021).

In 2015, the California State Legislature decreed it to be the "state amphibian" and it is now classed as a protected species in certain designated parks and regions, including the counties of Contra Costa, Alameda, and San Mateo (CDFW, 2014; Hammerson, 2008).

The American bullfrog (*Lithobates catesbeianus*, but also known as *Rana catesbeiana*), is to be found throughout the San Diego region, predominantly in the coastal wetlands. Although native to North America east of the Rockies, is an invasive species to the western United States and was first sighted during the late 1890s. It preys upon birds, bats, rodents, frogs (even its own young), snakes, turtles, and lizards and thus may have deleterious ecological effects (McKercher and Gregoire, 2020). It prefers to live in permanently inundated wetlands and so constructive habitat modification by man, such as converting these wetlands to be ephemeral, may help to control the San Diego region's bullfrog population.

The Baja California treefrog (*Pseudacris hypochondriaca hypochondriaca*) can either be grey or bright green in color, and lives in mountain of hill country streams, and is a very common frog seen in San Diego County. Interestingly, its familiar call is almost always heard in movies as nighttime background, even in areas where it is absent in reality. Its diet mainly consists of invertebrates, often flying insects (California Herps, 2021).

The Arroyo toad (*Anaxyrus californicus*) is found throughout the Pacific coastal regions from Monterey County in the north, to northwestern Baja California, inhabiting washes, arroyos, sandy riverbanks, and other riparian areas. Its diet includes a wide variety of invertebrates, but it predominantly feeds on trail-forming tree ants (California Herps, 2021).

Fish

Fish consist of many classes, such as jawless fish (Agnatha; lampreys and hagfish), cartilaginous fish (Chondrichthyes; sharks and rays), and bony fish (Osteichthyes; ray-finned fish and lobe-finned fish); phylogenetic studies have demonstrated a more complex classification system, which we will not elaborate upon here.

Although not as common as farther north, there have been salmon runs in the past in San Diego County on the Tijuana River as well as in the San Mateo Creek, which borders San Diego and Orange Counties (Mulvaney, 2013). The southern steelhead (*Oncorhynchus mykiss irideus*) is an anadromous (that is, a fish that migrates up rivers from the sea to spawn) coastal rainbow trout and is the southernmost anadromous salmonid in North America. Compared to their northern cousins, the Chinook salmon (*O. tshawytscha*), they apparently spend fewer years in the freshwater waterways, probably due to the inhospitable conditions often found in southern California stream (low flows, warm temperatures) and have adapted to living under highly variable condition. They are also found in lagoons, which prove a more persistent and reliable water environment (Moyle, Israel, and Purdy, 2008).

There are only a few other native fish that have been recorded in both the San Diego Bay and Mission Bay: the shortfin corvina (*Cynoscion parvipinnis*) has been found in the San Diego Bay, Mission Bay, and the Sweetwater River, the tidewater goby (*Eucyclogobius newberryi*) in the Santa Margarita River (north San Diego County) (Smith et al., 1993).

Pacific lamprey (*Entosphemus tridentatus*) has generally the same distribution in the San Diego region as the southern steelhead and are also anadromous; they are distinct from the non-parasitic Pacific brook lamprey (*Lampetra pacifica*) which has now extirpated from southern California (Smith et al., 1993).

Many other fish have been introduced into lakes for either recreational fishing (such as bass or catfish) or as bait (golden shiner or goldfish, for example); for reasons of brevity, we have not included a summary here.

Invertebrates

The word "invertebrate" is no longer used by biologists as a valid classification term and the organisms we will be describing below in fact are in many ways even more dissimilar to one another than mammals are to fish. The more correct terminology for these "animals without a backbone" organisms is "protostomes," which describes a particular sequence of gut formation during embryonic development and which we need not describe in more detail here. In contrast, the echinoderms and the vertebrates or chordates are referred to as "deuterostomes" (Martín-Durán et al., 2016). Therefore, although echinoderms are not protostomes, we will nevertheless include them in this section.

By almost any measure, the invertebrates are a fascinating group. In sheer numbers, there are more of them than any other animal group. They also have the greatest diversity of any group: they make up 95% of all animal species. Yet they are easily overlooked. They include insects, spiders, crabs, snails, clams, squids, octopuses, earthworms, leeches, jellyfishes, sea anemones, and many, many more. Most of these are represented in the San Diego region.

Invertebrates, whether aquatic, marine, or terrestrial, are important and often under-studied components of any habitat (Maffe, 2000). Few reports focus on them. Yet they serve as sensitive indicators of the overall health of the environment. Many invertebrates are important for other organisms. They act as pollinators, herbivores, scavengers, predators, and prey. Without the pollination of food crops by honeybees, human diets would be quite different. Unfortunately, the San Diego Bay and the Tijuana River estuary are some of those areas in which these important organisms have been under-studied.

We have arbitrarily separated them into two groups: terrestrial and marine invertebrates.

Terrestrial Invertebrates

Insects

The San Diego metropolitan region is a major urban area with more than 3 million people (www.statista.com). Amazingly, it is also a hot spot for insect diversity. The intersection of those sets of organisms also means that the San Diego region is a hot spot for threatened species: more than half of the arthropod species listed as endangered by the US government are in San Diego County (Dobson et al., 1997).

The insect diversity in the San Diego region results from two factors. First, California overall is a biological hotspot due to its geology, soil, and climate. Its latitude ensures a Mediterranean climate, and the elevation varies from sea level to the Sierras. It is split by the San Andreas fault so that part is on the Pacific plate and another part is on the North American plate. Second, the combination of the hills and bays yield a multitude of microenvironments with different climates and soils.

Unfortunately, much of the San Diego region has been developed. Buildings and streets cover more than 40% of the developed land, and the habitat for insects has been lost. For example, about 43% of the butterfly species have been lost. In addition to loss of habitat, those areas that can support insects have been fragmented. As is the case with many species of wild animals, the inability to move from one area to another is highly stressful. Also, the quality of those undeveloped areas does not allow them to support a sufficient population of some insects. Invasive species have also put pressure on native insects. These might be insects that compete with the native insects or even invasive plant species that crowd out native plants and reduce food sources for insects. Ice plant (*Carpobrotus edulis*), which is frequently used as easy to grow and low-maintenance groundcover, changes the nature of the soil that it grows in so that there are fewer insects and other invertebrates.

Preventing exotic insects from gaining a foothold is the best strategy for dealing with invaders. However, it is not always possible, and so, additional strategies are necessary to eradicate non-native insects (Liebhold and Kean, 2019). Many nascent infestations fail because the insects cannot establish themselves in the new environment. They might not find suitable food or habitat or there might be too few insects to succeed. However, once the non-native insects get a start, controlling them in a forest is very challenging. The good news is that not every insect must be killed. The population only needs to be reduced below a threshold.

The San Diego Natural History Museum has an excellent website that lists and describes many of the insects of the San Diego region (http://archive.sdnhm.org/field guide/index.html). Some insects are more welcome than others by the general public: bees, butterflies, and dragonflies.

Dragonflies and damselflies are a more primitive insect; their ancestors first appearing during the Carboniferous Period, around 325 mya (Resh and Cardé, 2009). They have two pairs of wings rather than the single pair of wings and the paired halteres or the paired elytras of more advanced flying insects (for example, flies or beetles). Some of the common dragonflies include common green darner (*Anax junius*), blue-eyed darner (*Rhionaeschna multicolor*), California darner (*Rhionaeschna californica*), spot-winged glider (*Pantala hymenaea*), red damselfly (*Telebasis salva*), and Mexican amberwing (*Perithemis intensa*).

Grasshoppers and crickets have been around for 250 million years. These ground-dwelling insects have powerful rear legs that allow them to escape predators. Those in the San Diego region include the California rose-winged grasshopper (*Dissosteira pictipennis*), the yellow grasshopper (*Schistocerca americana*), and the Jerusalem cricket (*Stenopelmatus fuscus*).

About 150 species of butterflies are found in the San Diego region. The famous monarch butterflies (*Danaus plexippus*) migrate through the area and settle on eucalyptus trees, a non-native species. Other species include the western tiger swallowtail (*Papilio rutulus*), western pygmy blue (*Brephidium exilis*), red admiral (*Vanessa atalanta rubria*), and painted lady (*Vanessa cardui*). Rarely seen are the two-tailed swallowtail (*Papilio multiculdata*), the variegated fritillary (*Euptoieta claudia*), the Mexican yellow (*Eurema mexicanum*), and Milbert's tortoiseshell (*Nymphalis milberti subpallida*); also rare is the Quino checkerspot (*Euphydryas editha quino*) and it is protected by the Endangered Species Act. The peninsular metalmark (*Apodemia mormo peninsularis*), Hermes copper (*Lycaena hermes*), and Thorne's hairstreak (*Callophyrus gryneus thornei*) are indigenous to the San Diego region; the latter is also rare.

Beetles are in the order Coleoptera. They are different from most other insects in that their front wings have hardened into wing cases called elytra. With 400,000 species, beetles are an amazing successful and diverse group of organisms. They account for about 40% of all insects and 25% of all known animals. They are well-represented in the San Diego region with many species, including the seven-spotted lady beetle (*Coccinella septempunctata*), spotted cucumber beetle (*Diabrotica undecimpunctata*), and cobalt milkweed beetle (*Chrysochus cobaltinus*).

The San Diego region has about 170 species of native bees from six families (Apidae, Andrenidae, Colletidae, Halictidae, Megachillidae, and Melittidae). Honeybees (*Apis mellifera*) are an introduced species, originally from Europe. Other common bees include the California bumblebee (*Bombus californicus*), yellow-faced bumblebee (B. *vosnesenskii* or *B. caliginosus*), California carpenter bee (*Xylocopa californica*), and green metallic bee (*Agapostemon texanus*). The European wool carder bee (*Anthidium manicatum*) is an introduced species that is spreading rapidly.

Wasps are narrow-waisted insects that can sting. There are tens of thousands of species. While some live in colonies, most are solitary. They are valuable pollinators. Several species occur in the San Diego region, including the European paper wasp (*Polistes dominula*), western yellowjacket (*Vespula pensylvanica*), blue mud wasp (*Chalybion californicum*), and tarantula hawk (*Pepsis pallidolimbata*).

The San Diego region used to be home to over 100 species of ants. However, the introduced Argentine ant (*Linepithema humile*) arrived in the San Diego region in the early 1900s and has driven out most of the native ants in urban areas. The Argentine ants are very small and often invade houses in search of food. In citrus groves and vineyards, the ants tend aphids that produce honeydew from the plants. They protect the aphids from other predators. They are detrimental to plants that depend on ants to disperse their seeds (Christian, 2001). Some have suggested that a supercolony of Argentine ants stretches from Northern California to the Mexican border, a distance of almost one thousand kilometers. However, genomic studies (of the ants show that they are different and probably the result of the introduction of multiple clones of the ants (Ingram and Gordon, 2003). Nevertheless, the ants do form massive colonies that include billions of ants and multiple queens (Moffett, 2012). Other ant species include the carpenter ant (*Camponotus semitestaceus*), orange carpenter ant (*Camponotus fragilis*), the acrobat ant (*Crematogaster* spp.), the harvester ant (*Pogonomyrmex californicus*), and red wood ant (*Formica integroides*).

These are only a few of the insects in the San Diego region. There are also praying mantises, walking sticks, and cicadas. However, in recent years, scientists have become worried about the decline in the number of insects. The cause of the decline seems to be human activities (Simmons, 2019). A series of papers has sounded the alarm. The loss of insects will affect our food supply. Hallmann et al. (2017) used traps across Germany to determine the number and types of insects. They found that flying insects have decreased by 75% over the 27 years of study. Lister and Garcia (2018) documented a similar decline in arthropods in the soil. Mathiasson and Rehan (2019) showed a loss of a significant number of bee species. Beekeepers noted that they lost 40% of their hives in the winter of 2018–19 (Neilson, 2019). The causes are thought to include decreasing crop diversity, poor beekeeping practices, loss of habitat, and indiscriminate use of pesticides. Infestation by a mite (*Varroa destructor*) has also harmed hives.

Arachnids

Arachnids are a large class of invertebrates that feature spiders, scorpions, ticks, mites, harvestmen, and more. Most have eight legs, but some have other appendages that might look like legs. In addition, there are some mites that have fewer legs. Unlike insects, they also have no antennae or wings, and they have two main body segments rather than the three of insects.

Spiders

Spiders are very common essentially everywhere, and the San Diego region has its share. The western black widow (*Latrodectus hesperus*) and the brown widow (*L. geometricus*) are one of the few spiders in the US that are venomous to humans. Fortunately, they tend to be shy and retiring. Tarantulas (*Aphonopelma reversum* and *A. eutylenum*) are large, hairy, and not harmful. The orbweaver (*Araneus bispinosus*) is a non-native species that builds large beautiful webs. The marbled cellar spider or daddy longlegs (*Holocnemus pluchei*) has a brown stripe on its ventral side and is also a non-native species. Another daddy longlegs is the longbodied cellar spider (*Pholcus phalangioides*); it is gray colored with an elongated abdomen.

Scorpions

The San Diego region has a number of scorpions (Buhler, 2018). They are rarely seen as they are nocturnal. They are also not dangerous unless the person stung happens to be allergic to the venom. Amazingly, they can most easily be found by using a black or ultraviolet light at night. The scorpions fluoresce brightly. They give birth to live young that ride on the mother's back early on. The most common species in the San Diego region is the bark scorpion (*Centruroides sculpturatus*). Others include the California common scorpion (*Paruroctonus silvestrii*), sawfinger scorpion (*Serradigitus gertschi*), and the California swollenstinger scorpion (*Anuroctonus pococki*).

Worms

We are most familiar with earthworms, but there are many, many others. Even earthworms are not well studied in the San Diego region, and many of those here are imports. The native earthworms can still be found in undisturbed areas. Central California and the Coastal Ranges has 22 species in 13 genera of five families (Acanthodrilidae, Lumbricidae, Megascolecidae, Ocnerodrilidae, and Sparganophilidae). Only six of the species are native to the area. The rest are invaders. Reynolds (2016) has an extensive list of all of the species and their locations.

Many worms are parasites, and some parasitize humans. Nematodes, which are roundworms, occur in many species of fish. They appear on the intestines, liver, in the body cavity, or in the flesh. They also live in seals, porpoises, whales, and dolphins. Their eggs escape in the feces of the mammal and are eaten by small crustaceans that are then eaten by fish or squid. The nematodes can also infect humans if they eat raw or unprocessed fish. Flukes and tapeworms are also common in the San Diego region fish and other animals.

Isopods (Woodlice)

These small crustaceans have segmented, dorso-ventrally flattened bodies with seven pairs of jointed legs. Females carry the fertilized eggs in their marsupium on the ventral side of its abdomen. The young emerge and seem to be a live birth, but they are really from eggs. There are thought to be 5–7,000 species worldwide. Most are useful in that they turn the soil like earthworms. However, they do eat some crops (for example, strawberries) and are considered a pest if they invade homes. One common group is the pill bugs or rolly-pollies (*Armadillidium vulgare)* that are familiar to most children. They are common under rocks, leaves, or boards in the backyard. They were introduced to the Americas from Europe, but have thrived here. They breathe through gill-like slits and so require moisture to survive.

Myriapods (Millipedes and Centipedes)

At first glance, millipedes and centipedes look like insects with lots of legs. However, they are quite different. Millipedes have two pairs of legs per body segment. While a small number are predators, most eat decaying matter. Centipedes have only one pair per segment. They are essentially carnivorous. They were also some of the earliest arthropods to colonize land during the Carboniferous Period (320 to 290 mya), the largest being *Arthroplura*, which could be at least 2 meters in length and 50 cm in width.

There are a number of species of each in the San Diego region (Shelley, 2002). The millipede *Tylobulus claremontus* is about 3–9 cm when mature. It is black to dull reddish-brown. They are valuable to the ecosystem because they breakdown plant matter (especially redwood litter) into humus. The greenhouse millipede (*Oxidus gracillis*) is an introduced species almost globally in both the tropics and temperate zones. It is about 2 cm in length and is often considered a pest in greenhouses.

There are four groups of centipedes living in the San Diego region. Stone centipedes (*Lithobiomorpha* spp.) are small (usually less than 1.5 cm) and have 15 pairs of legs as adults. They are common in gardens and eat small insects. Soil centipedes (*Geophilomorpha* spp.) small and can have more than 60 pairs of legs. They live in the ground and eat subterranean insects. House centipedes (*Scutigeromorpha coleoptrata*) include only one species. They have 15 pairs of long legs and are about 5 cm long. These are the only local centipedes that can bite, but their fangs are weak. The desert or tiger centipede (*Scolopendra polymorpha*) is known to inhabit San Diego and often invades homes. They are one of North America's largest centipede, frequently reaching a length of between ten to 17 centimeters; it is venomous and its bite can induce a painful reaction and very occasionally lead to fatal anaphylactic episodes (UCR News, 2015).

Gastropods (Snails and Slugs)

Snails and slugs are both mollusks. Slugs have lost the shell that covers snails. All produce mucus to facilitate their movement and protect against desiccation. For a more extensive list, the reader is referred to the *Checklist of the Land Snails and Slugs of California* (Roth and Sadeghian, 2006).

Slugs and snails have a radula, which they use for feeding. By eating leaves, animal droppings, moss, and dead plant material, they produce soil humus and provide a valuable service. Two common species in the San Diego region are the three-banded garden slug (*Lehmannia valentiana*) and the tramp slug (*Deroceras invadens*). Both are terrestrial slugs and feed on plants and decomposing wood; they are also considered pests as they enjoy orchids and ornamental plants (South, 1992; Baker, 1999)

The most common snail in the San Diego region is the non-native European garden snail (*Cornu aspersum*), but there are several native species (UCANR, 2021). The Monadenia land snail *(Monadenia fidelis leonina)* has a very dark shell and beautiful purple flesh. Montezuma shoulderband snails (*Helminthoglypta montezuma*) are the most common native land snails in the San Diego region and are easily confused with garden snails. They are found mostly in undisturbed native ecosystems. The Santa Catalina lancetooth *(Haplotrema catalinese*) eats plant material and other snails and slugs.

Marine Invertebrates

Many invertebrates are found in the San Diego and Mission Bays and the Pacific Ocean (NPS, 2018). The numbers of different types of organisms are almost beyond counting. They can be best seen while snorkeling, scuba diving, or just looking at tide pools. The tide pools at the La Jolla Cove and Torrey Pines State Reserve Beach, California, feature an extraordinary number of algae, crabs, mollusks, seastars, and others organisms. They are easily visible at low tide.

Here we will name a sampling of common species. For a more detailed account of marine invertebrates, we recommend this outstanding book *Intertidal Invertebrates of California* by Morris et al. (1980).

Porifera

Sponges

The bodies of sponges have pores that allow water to circulate. They are filter feeders and feed on single-celled organisms, detritus, and other material that they filter out of the water. The red beard sponge (*Clathria prolifera*) is a red or orange-brown sponge with many projections. This sponge was originally from the Atlantic Ocean and arrived in the Bay in the 1940s. The yellow sponge (*Halichondria bowerbanki*) appears as a flat mass on objects. It eats plankton. It was also from the Atlantic, but arrived in the Bay in the 1950s.

Cnidarians

Jellies

Jellies are graceful, beautiful, and sometimes dangerous. They go with the flow, literally. The current moves them along, but a few can also propel themselves to a degree. They vary in size from almost microscopic to specimens with tentacles of over 35 m. Among the many species found in the Bay and Pacific Ocean near the Bay are these. The moon jelly (*Aurelia* spp.) resembles a large dinner plate. It is translucent to whitish with short tentacles at the rim and four longer tentacles at the center. It eats plankton that is captured in the mucus on the tentacles and passed up to the mouth in the bell. The bell medusa (*Polyorchis* spp.) is globular and up to 4 cm tall. Tentacles hang from the rim. It eats zooplankton, which are transported to the mouth as in the moon jelly. Comb jellies (*Pleurobrachia* spp.) are transparent with eight rows of "combs" or cilia. It has two tentacles that collect food. These jellies bioluminesce in the dark. Their cilia aid in movement. The orange anemone (*Diadumene* spp.) is orange, flesh, or salmon pink and is also an introduced group.

Sea Anemones

Anemones look like underwater flowers (Gong, 2019). They have two lives: first, they form stationary polyps, and second, they produce eggs that become mobile planula larva. Their phylum Cnidaria also includes corals, sea pens, and gorgonians. All anemones are carnivorous. The common California genus *Anthopleura* eats anything that gets caught in its stinging tentacles. The orange-striped green anemone (*Diadumene lineata*) attaches to various underwater objects. It is shiny green, olive, or olive brown with orange stripes. This non-native anemone was introduced years ago to the Bay, and it is now worldwide.

Annelids

Flatworms

Flatworms belong to the phylum Platyhelminthes (Watkins, ND). Two of the three classes Cestoda (tapeworms) and Trematoda (flukes) are parasitic, and one, Turbellaria, is free-living. They are very simple and lack a respiratory or circulatory system. They absorb oxygen through their skin. Their mouth is at mid-body. Food enters the mouth and waste is expelled from the mouth also. The free-living flatworm *Notoplana acticola* is 2.5–7.6 cm long with two eyespots on the dorsal surface. It is brown or gray and eats small crustaceans, zooplankton, other worms, and dead animals. Flatworms are very colorful and graceful in the water.

Round Worms and Worms

Round worms belong to the phylum Annelida, and they are both terrestrial and marine. *Megasyllis nipponica* is originally from Japan (Tighe, 2019). They live mostly in the mud, rocks, and other places at the bottom of the Bay. In the summer, they change radically. Adult worms move eggs and sperm to the rear of the worm and bud off a new worm. This schizogamy yields bright orange epitokes that swim around looking for partners. Upon finding a partner, they burst open to allow the sperm to fertilize the eggs. The native scale worm (*Halosydna brevisetosa*) is about 5 cm long with 18 pairs of dorsal scales. It can be red, tan, or brown. They are scavengers. Some live in the tubes of tubeworms in a form of commensalism.

The leech *Branchellion lobata* is about an inch or so long with a sucker at each end. It feeds by attaching to sharks, fish and squids and drinking their blood. They begin as males and later become female.

Like many invertebrates, the worms can also be quite strange. The echiuran *Urechis caupo* (the fat innkeeper worm) inhabits a U-shaped burrow in both intertidal and subtidal mudflats. It "plugs" the burrow and uses its body to pump seawater through the burrow so that it can capture food (Arp et al., 1992). Other invertebrates live with the worm and feast on its leavings (Parr, 2019). These include the clam (*Cryptomya californica*), the scale-worm (*Hesperonoe adventor*), the pea crab (*Scleroplax granulata* or *Pinnixa franciscana*), and the hooded shrimp (*Betaeus longidactylus*). Strong storms can break up the sand in which the worms live, and occasionally, thousands of them can be seen stranded on beaches in Northern California.

Invasive species can travel in a surprising way. For example, bait worms (*Nereis virens*) are harvested in brown seaweed (*Ascophyllum nodosum*) from Maine and shipped many places, including the San Diego region. Once the worms are used, the boxes are often tossed into the water. Unfortunately, the boxes still contain other organisms. Haska et al. (2012) found 13 species of macroalgae and 23 species of invertebrates in the boxes. Two potentially toxic microalgae (*Alexandrium fundyense* Balech and *Pseudonitzschia* multiseries [Hasle]) were also found. *A. fundyense* was found in the San Diego region.

Mollusks

Many mollusks are sessile or creep slowly across a surface on their (usually) single foot; however, in the larval stage, they are free-swimming trochophores and which make up a large proportion of the zooplankton in the ocean surface. We will describe a number of mollusks that are typically found on the Southern California seashore.

Chitons

Chitons are a small to medium-sized mollusk having a single upper shell, made up of about eight separate segments, which is both flexible (for accessing odd-shaped spaces) and hard (for protection from predators). In California, the largest found is *Cryptochiton stelleri*, but this inhabits only the cooler Pacific waters north of the central coast. In southern California, there are about 36 species, with most living in the ocean depths of between 30 and 305 m. Usually spotted in rock depressions on the southern California beaches is the conspicuous chiton (*Stenoplax corrugata*), about 15 cm in length, as well as *Katharina tunicata*

(10–12 cm), *Mopalia muscosa* (8 cm), *Stenoplax heathiana* (8 cm), *Cyanoplax hartwegii* (5 cm) and *Nuttalina californica* (5 cm) (Stebbins and Eernisse, 2009; Watanabe, 2017).

Clams

Clams are bivalves. Their two calcareous shells are connected by two adductor muscles, and they have a power foot for burrowing into the sand. Unlike oysters and mussels, they do not attach to a substrate. The term "clam" is typically used to describe those bivalves that are edible. They are all filter feeders.

There many species of clam that inhabit the San Diego seashore. The Pismo clam (*Tivela stultorum*) is brown/grey, tan, or yellow and is one of the most common found on the beaches. The boring softshell clam (*Platyodon cancellatus*) is white or gray with a dark siphon and the small California sunset clam (*Gari californica*) is white with pink lines along the growth zones (iNaturalist ND).

Invertebrates, including clams, can also be very destructive. For centuries, the shipworm *Teredo navalis* was the bane of wooden ships and other wooden structures in the water. Although it is called a "worm," it is actually a clam.

Mussels

The California mussel *Mytilus californianus*, also a bivalve, is bluish grey. These filter-feeders eat detritus and microscopic plants and animals. They form dense clusters on rocks, piers, and other substrates that are in calm waters. The tiny bifurcate mussel *Mytilisepta bifurcata* is dark with white blending to pink-brown and brown bands (iNaturalist ND).

Oysters

Oysters are bivalve filter-feeders that eat plankton, bacteria, and detritus. They are found throughout the intertidal zone. The Olympia oyster (*Ostrea lurida*) is the common native oyster in the San Diego Bay. The shell is 5–8 cm long with a wavy edge. Recently the Pacific oyster (*Magallana gigas*), native to Japan but cultured locally, has become an invasive species; it has an important effect upon the environment as it builds reefs and this may change the shoreline ecosystem. It is also a threat to the native oyster restoration program (Johnson, 2013)

Oysters have been important in southern California for hundreds of years. The Native Americans used them as a handy source of food and built mounds with the left-over shells.

Abalone

The abalone is a relative giant among the gastropod mollusks and is classified a basal marine snail of the genus *Haliotis*; some species may live for at least 70 years (Haaker et al. 2001). In the eastern Pacific, they range in adult size between 15 and 30 cm (Hoiberg 1993). Abalone are characterized by having between three and seven open holes in their shells (respiratory pores) and a distinctive lining of mother-of-pearl (nacre) in their shells, thus contributing to their commercial and decorative desirability (Haaker et al. 2001). There are up to five species found off the coast of San Diego, each occupying a different habitat. The smallest is the white abalone (*H. sorenseni*) 20–25 cm, followed by the green abalone (*H. fulgens*) and the pink abalone (*H. corrugata*), both about 25 cm in diameter; the black abalone (*H. cracherodii*; 20 cm) and the red abalone (*H. rufescens*), the largest, are to be found throughout the Pacific Ocean from the coasts of central Oregon southwards to Baja California (Geiger 2012). Harvesting of abalone has always been undertaken by divers, never by net, but overfishing and disease has taken many species, such as the white abalone, to near-extinction by the mid-20th Century (Haaker et al. 2001). Commercial fishing was placed under a moratorium by the California Department of Fish and Wildlife after 1997 and recreational harvesting was closed for the Southern California coast from 1997 until 2026 at the earliest (Allen et al., 2005). Recently, marine biologists have established a breeding program for the endangered white abalone at the UC Davis Bodega Marine Laboratory in Bodega Bay, northern California (Duggan 2021).

Nudibranchs

The words "sea slugs" might conjure up a sea-going version of the slugs we see on land. Nothing could be further from the truth. Sea slugs are gastropods and come in many colors; many are festooned with cerata (soft projections) (Ueda and Agarwal, ND). As a result, common nudibranchs, sea hares, sapsucking slugs are some of the most beautiful of sea creatures. Over 3,000 species of nudibranchs are known (Bergamin, 2014). There are two forms of nudibranchs. Dorids are large, round, and flat. Aeolids are smaller with lush, feathery gills called cerata. The native Spanish shawl nudibranch (*Flabellinopsis iodinea*) has a combination of almost violet ventral side, a flashy pink upper side with a bright orange cerata. They eat hydroid that are pigmented with astaxanthin, giving the nudibranch its distinctive colors. Other nudibranchs native to the San Diego region is the spotted triopha (*Triopha maculata*) and the California aglaja (*Navanax inermis*) a predatory sea slug that feeds upon other gastropods such as *Bulla* and *Haminoea*. Sea hare (*Phyllaplysia taylori*) is a primitive marine gastropod lives primarily in eelgrass (*Zostera marina*) and feeds upon the sponge larvae and diatoms that settle on the eelgrass blades.

Cephalopods (Octopi and Squid)

Squid and octopi are, despite their appearance, also mollusks. They can change colors quickly. While octopi were common in the Bay in the 19th Century, today they are rare. Two species of octopi live in the Bay. With an arm span of over 60–90 cm is the giant Pacific octopus (*Enteroctopus dofleini*) which can grow to nearly 5 meters in length. Those octopi with an arm span less than 60–90 cm are either the East Pacific octopus (*Octopus rubesens*) or a young *E. dofleini*. The Humboldt squid (*Dosidicus gigas*) periodically arrives by the thousands in Northern California. They are about 7–27 kg. Most squid caught for eating are California market squid or opalescent inshore squid (*Doryteuthis opalescens*). They are found near shore and reach a length of about 25 cm.

Crustaceans

Crustaceans are a very large group of arthropods that includes crabs, lobsters, crayfish, shrimp, krill, woodlice, and barnacles. About 67,000 species are known. As with the other arthropods, they all have exoskeletons that must be shed for the animal to grow.

Crabs and Lobsters

Crabs are decapod crustaceans with a thick exoskeleton and a pair of pincers. Crabs live in both salt and fresh water around the world. Multiple species live in the San Diego region. The Dungeness crab (*Cancer magister*) is the best-known crab in the San Diego region. It is a favorite dish for residents and tourists alike. The San Diego Bay is a nursery for the Dungeness, and crabbing in the bay is illegal. The tuna crab (*Pleuroncodes planipes*), also called the pelagic red crab, is sometimes found washed up on the shore in huge numbers, probably due to warm water intrusion; it is in fact a species of squat lobster which normally live off Baja California (Aguilera, 2015). The native spider crab (*Pyromaia tuberculate*) eats algae and other plants. Sponges, algae, and other organisms grow on the crab's body and legs and give it camouflage.

Shrimp

Shrimp are decapod crustaceans with elongate abdomens. In this way, they are more similar to lobsters than crabs. Their many swimmerets allow them to swim well. There are thousands of species, and they are an important food crop.

The Bay contains multiple species. The Korean shrimp (*Palaemon macrodactylus*) is reddish with a prominent rostrum. These shrimp are omnivores and often scavenge. This non-native species arrived in the Bay in 1957. Bay shrimps (also called grass shrimp) (*Crangon* spp.) are semi-transparent with black spots. They eat smaller shrimp, amphipods, clams, and plants. This is a native species. They are caught now mostly for bait. While Native Americans likely enjoy the bay shrimp, early Spanish and Americans ignored them until Chinese workers began using them. Bay shrimp are sensitive to salinity and temperature. They like brackish water with a salt concentration of 14–24 parts per thousand and about 18 °C. They will migrate for miles to find this combination. They are protandrous hermaphrodites. That is, they begin life as males for a year and then change to females.

Snapping shrimp (family Alpheidae) create an audible sound that can be heard in the bay. In fact, it is second only to the echolocation clicks of sperm whales in volume. They use specialized claws to make the sounds, stun prey, and to communicate. The Bay ghost shrimp (*Neotrypaea californiensis*) is pale in color and grows to about 13 cm. It is a filter feeder that lives in a burrow and reworks the bay bottom somewhat like earthworms do for soil. They disturb oyster beds.

The San Diego fairy shrimp (*Branchinecta sandiegonensis*) is an endangered species and lives in small, shallow, vernal freshwater pools (Anonymous, 2019).

Echinoderms

Sea Stars

Sea stars (also known as starfish) are echinoderms of the superphylum Deuterostomia. About 1,500 species are known, and they are found around the world. They feature a central disc and usually five arms, but some species have many arms. Their colors vary. Although they seem peaceful, they are actually predators. The pink bay star (*Pisaster brevispinus*) is pink. It eats mussels, other bivalves, and sand dollars. It is found on the sea floor of the Bay and on pier pilings since it needs a high level of salinity. The brittle star has five thin very flexible arms. It is brown or gray and eats detritus and plankton.

Since 2013, sea stars have been suffering from a disease called sea star wasting disease (Jaffe et al., 2019). More than 20 species have been affected. The disease causes necrotic lesions, twisted rays, ray loss, and death. It seems to be viral, but the causative agent has not been identified.

Jaffe et al. (2019) examined the disease in *Leptasterias* spp. This small sea star broods its young rather than releasing them as planktonic larvae, and thus, they might be more susceptible to a viral disease. In 2010, this sea star was common in the Bay. In 2016, few could be found. The researchers found that this sea star suffered symptoms similar to larger sea stars. The loss of this species is not a good sign for the general health of the Bay. Twenty other species are also in decline.

Harvell et al. (2019) studied the loss of the common predatory sunflower sea star (*Pycnopodia helianthoides*) throughout its traditional range along the West Coast. Sea star wasting disease is responsible, and they found that its outbreaks were associated with unusually warm surface temperatures.

PLANTS

Introduction

As noted earlier, the San Diego region comprises areas of Mediterranean, semi-arid climates and even desert conditions, as well as marine areas, for many plants and fungi (Kottek et al., 2006). With this variety of habitats, the Bay region has a diverse array of plants. Here we will offer a representative sampling of those. For a more extensive listing of the plants of the region, we recommend the extensive catalog of vascular plants prepared by Rebman and Simpson (2014). The ornamental species planted by landscapers, businesses, and homeowners are too numerous to mention here, given that our aim was to describe the native environment. The reader should refer to other works that give more weight to such a topic.

Plants are classified into a number of groups. The most primitive are the mosses, liverworts, and hornworts, more advanced plants include ferns and their allies (such as glubmosses and horsetails), the most complex plants are the vascular forms, such as tree ferns, gymnosperms, and angiosperms (flowering plants) that include both dicotyledons and monocotyledons (Whittaker, 1969; Margulis, 1971).

FIGURE 8.2 Wetlands. Wetlands once surrounded San Diego Bay and other estuaries in the region. Over the last 150 years, much of the wetlands have been lost to filling and development. However, the wetlands are critical for flood protection and more. Recently, efforts have been underway to restore wetlands (Photograph reproduced with permission from Sebastian Kaser).

Plants Associated with Water

San Diego Bay was long surrounded by marshland, wetland, and tidal flats that were regularly inundated by tidal action (Figure 8.2). However, much of the low-lying shoreline has since been developed for housing and commercial use. This has resulted in a transition from the once-native pickleweed (*Salicornia europaea* and *S. virginica*) to the invasive cordgrass (*Spartina*), the more salt-tolerant native salt grass (*Distichlis*), and the invasive Mediterranean saltwort (*Salsola soda*) and Australian saltbush (*Atriplex semibaccata*) (Baye, 2006).

Eelgrass (*Zostera marina*) provides a home to Taylor's sea hare (*Phyllaplysia taylori*), a type of sea slug gastropod mollusk (Nudibranch), and a spawning habitat for the Pacific herring (*Clupea haengus pallasi*) and outmigrating juvenile salmon (*Oncorhyncus spp.*) (Ort et al., 2012). Below the surface, at an average depth of about four meters, the soil is rich in carbon, providing a large storage system of carbon in the San Diego Bay (Poppe and Rybcyk, 2018).

The La Jolla kelp forest features an amazing array of organisms. One of the larger organisms is the giant kelp (*Macrocystis* spp.), a marine macroalga whose photosynthesis provides nourishment to a large ecosystem. The kelp forest comprises stripes (stalks) and fronds (leaves) up to 200 meters in height and which provide shelter to hundreds of species (Foster and Schiel, 1985). Under optimal conditions, they may grow at a rate of 30–60 cm per day (National Ocean Service, 2013; 2020).

Native Land Plants

Native plants, such as the Montara manzanita (*Arctostaphylos montaraensis*) or the Antioch Dunes evening primrose (*Oenthera deltoides howellii*) are listed as critically endangered species, mainly due to land-use change, new developments, and off-trail/road walking and vehicle (for example, motorcycles, mountain bikes) habitat degradation (California Native Plant Soc., 2017).

Trees

The oaks are important trees in California. Their acorns were a large source of carbohydrate and protein for Native American tribes. California live oak (*Quercus agrifolia*), California black oak (*Quercus kelloggii*), canyon live oak (*Quercus chrysolepis*), interior live oak (*Quercus wislizenii*), and Engelmann oak (*Quercus engelmanni*).

The pines and other conifers (gymnosperms) are evergreen and lack flowers. They produce seeds in cones. They evolved about 300 million years ago. Several can be found in the San Diego region, including the bigcone Douglas-fir (*Pseudotsuga macrocarpa*), incense cedar (*Calocedrus decurrens*), and the western juniper (*Juniperus occidentalis*). Pines vary in size from shrubs to very tall trees. They evolved about 200 million years ago. They are native to the northern hemisphere with only a few examples in the tropics of the southern hemisphere. Pines hybridize easily, and this characteristic complicates their evolutionary history. The San Diego region has a number of different species, including the Coulter pine (*Pinus coulteri*), knobcone pine (*Pinus attenuata*), ponderosa pine (*Pinus ponderosa*), Jeffrey pine (*Pinus jeffreyi*), parry pinyon (*Pinus quadrifolia*), sugar pine (*Pinus lambertiana*), Torrey pine (*Pinus torreyana*), and single-leaf pinyon pine (*Pinus monophylla*).

Some trees tend to live in habitats near rivers, such as the California sycamore (*Platanus racemosa*), white alder (*Alnus rhombifolia*), and black cottonwood (*Populus trichocarpa*).

Another native tree is the California buckeye (*Aesculus californica*), also called the California horse chestnut, for its chestnut-like conkers, and which is found in the cooler coastal and foothill environments of Californian; it is drought and salt tolerant (SelecTree, 2020). Other trees and tree-like shrubs include the California bay (*Umbellularia californica*), madrone (*Arbutus menziesii*), bigleaf maple (*Acer macrophyllum*), western blue elderberry (*Sambucus nigra*), western redbud (*Cercis occidentalis*), and California black walnut (*Juglans californica*).

One significant botanical resident that has become almost endemic to the San Diego Bay area are the many different types of eucalyptus species (*Eucalyptus* spp., in particular the blue gum, *E. globulus*, and *E. camaldulensis*) that dot the hillsides in large groves; these are an invasive species that had been introduced from Australia to the San Diego region in the 19th Century with the expectation that they would provide a rapidly-growing source of timber and fiber for paper as well as for ornamental uses (Santos, 1997; Groenendaal, 1983). By 1875, Frank and Warren Kimball who founded National City, had planted eucalyptus. They later used the wood to fire kilns for bricks for a train station and planted many more trees along the Sweetwater River. However, the wood is so dense that harvesting and lumbering are extremely energy-intensive and thus they were left to their own devices. Efforts are underway to clear at least some of the most prolific and old stands (Wolf, 2015).

Shrubs

In addition to trees, a number of shrubs are also found in the San Diego region. For example, manzanita (*Arctostaphylos* spp.) are a number of species of shrubs and small trees in the chaparral regions of San Diego. The word means "little apple" in Spanish. They do well in poor soil and need little water. The bark is red or orange, and the branches are twisted. The berries and flowers of most are edible. Coyote bush (*Artemisia californica*) is a very common plant throughout much of California. It is a secondary pioneer plant in communities, such as coastal sage scrub and chaparral. It does not do well in shade and is easily replaced by other species that grow taller. Many species of sage (*Salvia*) occur in the San Diego region. In addition, hybridization is common. Some sage plants have trichomes (hairs) on the leaves that release an oil with a scent that makes them undesirable to animals and insects. Pacific poison oak (*Toxicodendron diversilobum*) contains an oil called urushiol that causes an allergic reaction in many people. Unfortunately poison oak is widespread throughout the region. Other shrubs include chamise or greasewood (*Adenostoma fasciculatum*), service-berry (*Amelanchier utahensis*), California sagebrush, desert willow (*Chilopsis linearis*), creosote bush (*Larrea tridentata*), snowberry (*Symphoricarpos mollis*), huckleberry (*Vaccinium ovatum*), skunkbush sumac (*Rhus trilobata*), nettle (*Urtica holosericea*), and flannelbush (*Fremontodendron californicum*).

Flowering Plants

Many flowering plants are found in the San Diego Bay region. For example, the state flower, the California poppy (*Eschscholzia californica*), is a common plant with beautiful orange flowers. Narrowleaf milkweed (*Asclepias fascicularis*) is a perennial with lavender or whitish flowers. The seedpods split open to release seeds with silky hairs. These plants are important larval host plants for Monarch butterflies. San Diego sunflowers (*Viguiera laciniata*) bloom much of the year. They are found in dry mesas, canyons and mountains in areas rich in chaparral and coastal sage scrub. Other flowering plants include the desert willow (*Chilopsis linearis*), laurel sumac (*Malosma laurina),* showy penstemon

(*Penstemon spectabilis*), scarlet bugler (*Penstemon centranthifolius*), lemonade berry (*Rhus integrifolia*), sugarbush (*Rhus ovata*), desert mallow (*Sphaeralcea ambigua*), and woolly blue curls (*Trichostema lanatum*). However, the list of flowering plants is enormous, and even the desert areas feature many flowers, especially in the spring.

Grasses and Sedges

Grasses and sedges (monocotyledons) are the most important food crop in the world. In addition, they are used for forage and decorative plants. They are found in many habitats and are the most common plant in the world. The San Diego region features multiple types of native grasses, including purple three-awn (*Aristida purpurea*), California fescue (*Festuca californica*), Idaho fescue (*Festuca idahoensis*), red fescue (*Festuca rubra*), Junegrass (*Koeleria macrantha*), giant wildrye (*Leymus condensatus*), California melic (*Melica californica*), buckwheats (*Eriogonum fasciculatum*), deer grass (*Muhlenbergia rigens*), purple needlegrass (*Nassella pulchra*), Indian ricegrass (*Oryzopsis hymenoides*), and pine bluegrass (*Poa secunda*). A number of grass-like plants are also important. They include the sedges (*Carex* spp.), rushes (*Juncus* spp.), western blue-eyed grass (*Sisyrinchium bellum*), yellow-eyed-grass (*Sisyrinchium californicum*).

Desert Plants

Many plants have adapted to living in arid desert conditions and are important habitat of insects and birds. The California juniper (*Juniperus californica*) is a shrub that grows to 3–8 meters. Its bark is gray and the foliage is bluish gray and scale-like. It provides food and shelter for turkeys, deer, and others. Native Americans also ate the berries. The fronds of the California fan palm (*Washingtonia filifera*) are 4 meters long with long, thread-like, white fibers. Natural oases with these plants provide shelter for many bird species and mainly occur downstream from hot springs. The creosote bush (*Larrea tridentata*) is an evergreen shrub of 1–3 meters. Their name derives from the characteristic smell that they emit. The jojoba (*Simmondsia chinensis*) is now grown commercially to produce jojoba oil. In nature, it provides food for deer, javelina, bighorn sheep, and livestock. Its nuts are eaten by squirrels, rabbits, and more. Other desert plants include the single-leaf pinyon (*Pinus monophylla*), Fremont cottonwood (*Populus fremontii*), Indian mallow (*Abutilon palmeri*), Mojave yucca (*Yucca schidigera*), Rush milkweed (*Asclepias subulata*), and sacred datura (*Datura wrightii*).

Ferns

The ferns are vascular plants that reproduce by using spores. They lack flowers and seeds. Their leaves are complex, and most produce the fiddleheads that are widely recognized. Fern occupy a wide range of habitats, but they tend to occur in those areas in which flowering plants are limited, such as shady areas, rock faces, and acid wetlands. Some are considered weeds, and others grow as epiphytes. Common examples in the San Diego region include the polypody ferns (*Polypodium californicum*), native sword ferns (*Polystichum munitum*), giant chain fern (*Woodwardia fimbriata*), goldback ferns (*Pteridium* spp.), wood ferns (*Dryopteris arguta*), and maidenhair ferns (*Adiantum jordanii*).

Non-vascular Plants

The San Diego region also includes quite a number of non-vascular plants. They are somewhat easy to overlook since the vascular plants now dominate most habitats. However, the non-vascular plants can still be found in various areas, including road cuts, rock outcrops, decaying logs, stream banks, and more. Others live as epiphytes in trees. The nonvascular plants form three divisions: mosses (Bryophyta), liverworts (Hepatophyta), and hornworts (Anthocerotophyta). They all lack vascular tissue and organs (for example, stems, roots, leaves, flowers). Some have leaf-like structures, but they are not true leaves. In place of roots, they have rhizoids that lightly anchor them into the soil. They include generalized cells called parenchyma, growth occurs in the meristem, and they are covered by a cuticle. The gametophyte is the dominant stage. The mosses (Bryophyta) in the San Diego region include *Tortula obtusifolia, Timmiella anomala, and Scleropodium obtusifolium* and the ashy spike-moss (*Selaginella cinerascens*). The liverworts (Hepatophyata) are represented by *Aneura pinguis and Asterella californica.* Examples of hornworts in the San Diego region include *Sphaerocarpos drewiae*, *S. michelii*, *S. texanus*, and *S. cristatus*.

Invasive Species

Invasive species can have a detrimental effect on native species. The movement of humans from continent to continent provides an excellent opportunity for plants and animals to move from one region to another. For example, the Shot Hole Borer is a boring beetle that drills into tree trunks and branches, bringing with it a pathogenic fungus along with other fungi that are conducive to establishing and nurturing Shot Hole Borer colonies. Shot Hole Borers are known to have attacked more than 200 species of native, exotic, and agricultural trees in Southern California and have been found in a number of environments—from urban landscapes to commercial groves, and now native riparian habitats like those within the Tijuana River Valley (SDPARKS ND).

Plants can also be invaders. For example, several species of invasive cordgrasses (*Spartina* spp.) are found in the San Francisco Bay, and they may compete with native species *S. foliosa*. *S. alterniflora* and *S. foliosa* easily form hybrids, and they are found mainly in the South and Central Bay (Ayres et al., 2004). They spread by rhizomes, and the hollow stems of smooth cord grass (*S. alterniflora*) grow from 0.6–1.3 meters tall. It was introduced to the West Coast in an attempt to prevent erosion, but it is now considered an invasive species. *S. anglica*, *S. densiflora, and S. patens* have had a more limited spread in the Bay. The exotic species outcompete the native species. They also change the mudflats to meadows and destroy the habitat of endangered species, such as the salt marsh harvest mouse. Moreover, the exotic species are detrimental to migrating shorebirds. Fortunately, the invasive cordgrasses have not yet reached as far south as the San Diego Bay.

FUNGI

The San Diego region has a wide array of fungi. They belong to two major groups. The basidiomycetes include common edible mushrooms. The fleshy structure that we eat is the fruiting body that is called the basidium. The main body of the fungi exists underground as a system of mycelium. The basidiomycetes include the common *Agaricus* of pizza fame, boletes, puffballs, and polypores. The other main group is the Ascomycetes, which form a structure called an ascus. These include morels, truffles, and cup fungi.

Common fungi of the San Diego region include the pleated pluteus (*Pluteus longistriatus*), garland roundhead mushroom (*Stropharia coronilla*), deadly skullcap (*Galerina marginata*), russet toughshank (*Gymnopus dryophilus*), arched earthstar (*Geastrum fornicatum*), yellow brain fungus (*Tremella aurantia*), split gill fungus (*Schizophyllum commune*), yellow staining mushroom (*Agaricus xanthodermis*), and lilac bonnet (*Mycena pura*).

REFERENCES

Abadía-Cardoso A, Freimer NB, Deiner K, Garza JC (2017) Molecular population genetics of the Northern Elephant Seal *Mirounga angustirostris*. *Journal of Heredity* 108(6): 618–627.

Ackerman JT, Peterson SH (2017) California gull diet, movements, and use of landfills in San Francisco Bay. *Tideline* 40: 1–2.

Aguilera M (2015) Red crabs invade San Diego shores, June 12, 2015, Scripps-UCSD, Scripps Institution of Oceanography, La Jolla, California, https://scripps.ucsd.edu/news/red-crabs-invade-san-diego-shores; accessed February 26, 2021.

Alexandrino J, Baird SJE, Lawson L, Macey JR, Moritz C, Wake DB (2005) Strong selection against hybrids at a hybrid one in the *Ensatina* ring species complex and its evolutionary implications. *Evolution* 59: 1334–1347.

Allen, B, Callaway, M, Haaker, P, Kalvass, P, Karpov, K, et al. (2005) Abalone recovery and management plan. California Fish and Game Commission, California Department of Fish and Game, Sacramento, California.

Anonymous. (2019), https://www.encyclopedia.com/environment/science-magazines/san-diego-fairy-shrimp; accessed February 26, 2021.San Diego fairy shrimp, encyclopedia.com entry

Arp AJ, Hansen BM, Julian D (1992) Burrow environment and coelomic fluid characteristics of the echiuran worm *Urechis caupo* from populations at three sites in northern California. *Marine Biology* 113: 613–623.

Audubon (ND) Guide to North American Birds, Bald Eagle, www.audubon.org/field-guide/bird/bald-eagle; accessed March 8, 2021.

Audubon California (ND) California snowy plover. https://ca.audubon.org/westernsnowyplover; accessed September 28, 2020.

Ayres DR, Smith DL, Zaremba K, Klohr S, Strong DR (2004) Spread of exotic cordgrasses and hybrids (Spartina sp.) in the tidal marshes of San Francisco Bay, California, USA. *Biological Invasions* 6: 221–231.

Baker GM (1999) *Naturalised Terrestrial Stylommatopha (Mullusca: Gastropoda),* Fauna of New Zealand, Number 38, Manaaki Whenua Press, Canterbury, New Zealand, pp. 1–254.

Barrat J (2013) Suburban raccoons more social yet dominance behavior remains that of a solitary animal. *Smithsonian Insider.* https://insider.si.edu/2013/07/suburban-life-does-not-alter-solitary-ways-of-the-raccoon/; accessed September 27, 2020.

Baye PR (2006) Selected Tidal Marsh Plant Species of the San Francisco Estuary: A Field Identification Guide, San Francisco Estuary Invasive Spartina Project (California Coastal Conservancy). www.spartina.org.

Bergamin A (2014) Nudibranchs, kings of the tidepool, command an audience. *Bay Nature.* https://baynature.org/article/nudibranchs-kings-tidepool/.

Beschta RL, Ripple RJ (2009) Large predators and trophic cascades in terrestrial ecosystems of the western United States. *Biological Conservation* 142: 2401–2414.

BirdLife International (2007) Species factsheet: California Condor *Gymnogyps californianus.* Downloaded from http://www.birdlife.org on 09/06/2020.

BirdLife International (2016) *Rallus obsoletus*, IUCN Red List of Threatened Species; Downloaded from http://www.birdlife.org accessed September 15, 2020.

Bochenski ZM, Campbell KE Jr (2006) The Extinct California Turkey, *Meleagris californica,* from Rancho La Brea: Comparative Osteology and Systematics. *Natural History Museum of Los Angeles County* (509).

Brehme CS, Hathaway SA, Booth R, Smith BH, and Fisher RN (2014) Research results of American badgers in western San Diego County, 2014, USGS Western Ecological Research Center, San Diego Field Station, San Diego, California.

Brown, CW, Stebbins, RC, (1964) Evidence for hybridization between the blotched and unblotched subspecies of the salamander *Ensatina eschscholtzii. Evolution* 18: 706–707.

Buhler B (2018) There are so many scorpions. *Bay Nature*, Winter 2018. Retrieved from https://baynature.org/article/there-are-so-many-scorpions/.

California Department of Fish and Game (2005) *Guide to Hunting Wild Turkeys in California.* California Department of Fish and Game, Wildlife Programs Branch: Sacramento, 42 pages.

California Herps (2020a) *Gartersnake, Thamnophis sirtalis. A Guide to the Amphibians and Reptiles of California.* www.californiaherps.com.

California Herps (2020b) Lizards. *A Guide to the Amphibians and Reptiles of California.* http://www.californiaherps.com/lizards/lizardspics.html.

California Herps (2021) A guide to the Amphibians and reptiles of California, San Diego gophersnake, www.californiaherps.com/snakes/pages/p.c.annectens.html; accessed March 9, 2021.

California Native Plant Soc. (2017) California Native Plant Society Inventory of Rare and Endangered Plants of California; California Native Plant Society, Rare Plant Program. 2017; *Inventory of Rare and Endangered Plants of California* (online edition, v8-03 0.39).

Carlisle, JD, Skagen, SK, Kus, BE, Riper, CV, Paxtons, KL, and Kelley, JF (2009) Landbird migration in the American west: recent progress and future research directions. *The Condor: Ornithological Applications* 111: 211–225, DOI: 10.1525/cond.2009.080096.

CDFW (2014) California red-legged frog named state amphibian. *California Department of Fish and Wildlife*. https://cdfgnews.wordpress.com/2014/07/15/california-red-legged-frog-named-state-amphibian/.

CDFW (ND-a) Keep me wild: Bobcat. *California Department of Fish and Wildlife*, https://wildlife.ca.gov/Keep-Me-Wild/Bobcat.

CDFW (ND-b) California condor. *California Department of Fish and Wildlife*. https://wildlife.ca.gov/Conservation/Birds/California-Condor.

Christian CE (2001) Consequences of a biological invasion reveal the importance of mutualism for plant communities. *Nature* 413: 635–639.

Cifelli RL (2000) Cretaceous mammals of Asia and North America. *Paleontological Society of Korea Special Publication* 4: 49–84.

Conroy CJ, Rowe KC, Rowe KMC, Kamath PL, Aplin KP, Hui L, James DK, Moritz C, Patton JL (2012) Cryptic genetic diversity in *Rattus* of the San Francisco Bay region, California. *Biological Invasions* 15: 741–758.

Cronin MA, Armstrup SC, Garner ER, Vyse GW, Vyse ER (1991) Interspecific and intraspecific mitochondrial DNA variation in North American bears (Ursus). *Canadian Journal of Zoology* 69: 2985–2992.

Devitt TJ, Baird SJ, Moritz C (2011). Asymmetric reproductive isolation between terminal forms of the salamander ring species Ensatina eschscholtzii revealed by fine-scale genetic analysis of a hybrid zone. *BMC Evolutionary Biology*, 11: 245. 10.1186/1471-2148-11-245.

DMT (2012) Area coyote, mountain lion sightings reported recently, Del Mar Times, December 13, 2012, Staff writer(s), Del Mar, California, https://www.delmartimes.net/sddmt-area-coyote-mountain-lion-sightings-reported-2012dec13-story.html; accessed March 5, 2021.

Dobson AP, Rodriguez JP, Roberts WM, Wilcove DS (1997) Geographic distribution of endangered species in the United States. *Science* 275: 550–553.

Dragoo JW, Honeycutt RL (1997) Systematics of mustelid-like carnivores. *Journal of Mammalogy* 8: 426–443.

Duggan T (2021) Bodega lab a key part of effort to save abalone, *San Francisco Chronicle*, San Francisco, California, April 6, 2021, pp. A1–A9.

Economist (2012) Commercial whaling: Good whale hunting. *The Economist*, March 4, 2012.

Edvenson JC (1994) Predator control and regulated killing: A biodiversity analysis. *UCLA Journal of Environmental Law and Policy* 13: 31–86.

Federal Records (2020) San Diego Climate Graphs, National Weather Service, NOAA, San Diego Climate Graphs (weather.gov), San Diego, CA; accessed February 27, 2021.

Fedriani, JM, Fuller, TK, Sauvajot, RM, York, EC (2000). Competition and intraguild predation among three sympatric carnivores. *Oecologia* 125(2): 258–270.

Feldhamer GA, Thompson BC, Chapman JA (2003) *Wild Mammals of North America: Biology, Management, and Conservation.* Johns Hopkins University Press, Charles Village, Baltimore, MD, p. 683.

Flynn JJ, Finarelli JA, Zehr S, Hsu J, Nedbal MA (2005) Molecular phylogeny of the Carnivora (Mammalia): Assessing the impact of increased sampling on resolving enigmatic relationships. *Systematic Biology* 54: 317–337.

Foster MS, Schiel DR (1985) The ecology of giant kelp forests in California: a community profile. *US Fish and Wildlife Service Report* 85: 1–152.

FS (ND) Forest Service, San Bernardino National Forest, www.fs.usda.gov/detail/sbnf/home/?cid=STELPRD3829099; accessed March 7, 2021.

FWS (2007) California condor, (Gymnogyps californianus). *U.S. Fish and Wildlife Service*. accessed September 28, 2020.

FWS (2020) California condor population information. *US Fish and Wildlife Service*. https://www.fws.gov/cno/es/CalCondor/Condor-population.html.

Geiger, DL, Owen, B (2012) *Abalone: Worldwide Haliotidae*. Conchbooks, Hackenhiem, Germany.

Gong AJ (2019) Voracious flowers of the tidepool. *Bay Nature*. https://baynature.org/2019/08/13/voracious-flowers-of-the-tidepool/.

Gould SJ (1977) *Ontogeny and Phylogeny*. Belknap Press, Cambridge, Massachusetts.

Graves CL (1964) An early San Diego physician: David Hoffman. *The Journal of San Diego History* 10(3); accessed 2010-07-03.

Grenfell WE Jr. (1974). *Food Habits of the River Otter in Suisun Marsh, Central California*. California State University, Sacramento. http://csus-dspace.calstate.edu/bitstream/handle/10211.9/1554/1974-Grenfell.pdf?sequence=1.

Groenendaal GM (1983) *Eucalyptus* helped solve a timber problem. In: *Proceedings of the Workshop on Eucalyptus in California*, pp. 1853–1880, Sacramento, California.

Haaker, PL, Karpov, K, Rogers-Bennett, L, Taniguchi, I, Friedman, CS, & Tegner, MJ (2001) Abalone, California's living marine resources: A status report. California Department of Fish and Game, Sacramento, California, pp. 89–97.

Hairston NG, Smith FE, Slobodkin LB (1960) Community structure, population control and competition. *American Naturalist* 94: 421–425.

Hallmann CA, Sorg M, Jongejans E, Siepel H, Hofland N, Schwan H, Stenmans W, Müller A, Sumser H, Hörren T, Goulson D, de Kroon H (2017) More than 75% decline over 27 years in total flying insect biomass in protected areas. *PLoS ONE* 12(10): e0185809.

Hammerson G (2008) *Rana draytonii. IUCN Red List of Threatened Species*. IUCN. 2008: e.T136113A4240307. doi:10.2305/IUCN.UK.2008.RLTS.T136113A4240307.en; http://www.fws.gov/sacramento/es/maps/CRF_fCH_FR_maps/crf_fCH_units.htm.

Harvell CD, Montecino-Latorre D, Caldwell JM, Burt JM, Bosley K, Keller A, Heron SF, Salomon AK, Lee L, Pontier O, Pattengill-Semmens C, Gaydos JK (2019) Disease epidemic and a marine heat wave are associated with the continental-scale collapse of a pivotal predator (Pycnopodia helianthoides). *Science Advances* 5(1): eaau7042.

Haska CL, Yarish C, Kraemer G, Blaschik N, Whitlatch R, Zhang H, Lin S (2012) Bait worm packaging as a potential vector of invasive species. *Biological invasions* 14: 481–493.

Heil J (2021) The Return of Red-legged Frogs, US Fish & Wildlife Service; Pacific Southwest region, Sacramento, California, www.fws.gov/cno/newsroom/Highlights/2021/Red-Legged-Frog/; accessed March 10, 2021.

Hensley AL (1946) A progress report on beaver management in California. *California Fish and Game* 32(2): 88.

Hertzog LA (1990) *Where North Meets South: Cities, Space, and Politics on the United States-Mexico border*. University of Texas Press, Austin, Texas, pp. 194–201.

Hoiberg, DH (1993) *Encyclopædia Britannica*. 15 Edition (ed Hoiberg DH). Chicago, IL: Encyclopædia Britannica, Inc.

iNaturalist (ND) Cardiff State Beach California, https://www.inaturalist.org/places/cardiff-state-beach#page=3&taxon=47115; accessed February 26, 2021.

Ingram KK, Gordon DM (2003) Genetic analysis of dispersal dynamics in an invading population of Argentine ants. *Ecology* 84: 2832–2842.

Jackman TR, Wake DB (1994). Evolutionary and historical analysis of protein variation in the blotched forms of salamanders of the Ensatina complex (Amphibia: Plethodontidae). *Evolution* 48: 876–897.

Jaffe N, Eberl R, Bucholz J, Cohen CS (2019) Sea star wasting disease demography and etiology in the brooding sea star *Leptasterias* spp. *PLoS ONE* 14(11): e0225248.

Jennings MR (1983) *Masticophis lateralis* (Hallowell), Striped racer. Catalogue of American Amphibians and Reptiles. *Society for the Study of Amphibians and Reptiles*; University of Texas, Austin, Texas, 343: 1–2.

Jiang D, Klaus S, Zhang Y-P, Hillis DM, Li J-T (2019) Asymmetric biotic exchange across the Bering land bridge between Eurasia and North America. *National Science Review* 6: 739–745.

Johnson CS (2013) *A New Oyster Invades, Sea Grant California*. University of California, San Diego, La Jolla, California, https://caseagrant.ucsd.edu/news/a-new-oyster-invades; accessed February 26, 2021.

Johnson WE, O'Brien SJ (1997) Phylogenetic reconstruction of the Felidae using 16S rRNA and NADH-5 mitochondrial genes. *Journal of Molecular Evolution*, 44: (suppl 1): S98–S116.

Kemp TS (1987) Fossil synapsids: The ecology and biology of mammal-like reptiles. *Science* 236(4803): 862–863.

Kemp TS (2005). *The Origin and Evolution of Mammals*. Oxford University Press, Oxford.

Kenyon KW (1969) *The Sea Otter in the Eastern Pacific Ocean*. U.S. Bureau of Sport Fisheries and Wildlife, Washington, D.C.

King JL (2004) The Current Distribution of the Introduced Fox Squirrel (*Sciurus niger*) in the Greater Los Angeles Metropolitan Area and its Behavioral Interaction with the Native Western Gray Squirrel (*Sciurus griseus*) Master's thesis, California State University, Los Angeles.

Klymkowsky MW, Cooper MM, Begovic E, Lymkowsky R (2016) Sexual dimorphism. *Biology LibreTexts*. https://bio.libretexts.org/@go/page/4120; accessed September 27, 2020.

Kottek M, Grieser J, Beck C, Rudolf B, Rubel F (2006) World Map of the Köppen-Geiger climate classification updated. *Meteorologische Zeitschrift* 15(3): 259–263. doi:10.1127/0941-2948/2006/0130; accessed January 3, 2021.

Krause WJ, Krause WA (2006) *The Opossum: Its Amazing Story* Archived 2012-12-11 at the Wayback Machine. Department of Pathology and Anatomical Sciences, School of Medicine, University of Missouri, Columbia, Missouri, p. 39.

Kucera T (1997) California Giant Salamander (Report). California Department of Fish and Game.

Kucher K (2010) Pond turtles surprise caretakers at wildlife refuge, San Diego Union-Tribune, 2010-06-01, San Diego, California; March 10, 2021.

Kutcha SR, Krakauer AH, Sinervo B (2008) Why does the yellow-eyed ensatina have yellow eyes? Batesian mimicry of Pacific newts (Genus *Taricha*) by the Salamander *Ensatina eschscholtzii xanthoptica*. *Evolution*, 62(4): 984–990. 10.1111/j.1558-5646.2008.00338.x.

Kuchta SR, Parks DS, Lochridge Mueller R, Wake DR (2009) Closing the ring: historical biogeography of the salamander ring species Ensatina schscholtzii. *Journal of Biogeography* 36: 982–995.

Lafferty KD, Tinker MT (2014) Sea Otters are recolonizing southern California in fits and starts. *Ecosphere* 5(5): 1–11, DOI: 10.1890/ES13-00394.1.

Lantz DE (1909) The brown rat in the United States. *US Department of Agriculture, Biological Survey* 33: 1–54.

Larsen DN (1984). Feeding habits of river otters in coastal southeastern Alaska. *Journal of Wildlife Management* 48: 1446–1452.

Liebhold AM, Kean JM (2019) Eradication and containment of non-native forest insects: Successes and failures. *Journal of Pest Science* 92: 83–91.

Lister BC, Garcia A (2018) Climate-driven declines in arthropod abundance restructure a rainforest food web. *Proceedings of the National Academy of Sciences USA* 115: E10397–E10406.

Litalien R (2004) *Champlain: The Birth of French America.* (Eds. Litalien, R, Vaugeois, D.) transl. Roth, K, McGill-Queen's Press, Montreal, Canada, pp. 312–314.

Leopold A (1949) Thinking like a mountain. In: *A Sand County Almanac: And Sketches Here and There*. Oxford University Press, Oxford, UK.

Maffe WA (2000) A note on invertebrate populations of the San Francisco Estuary. In: Goals Project. 2000. *Baylands Ecosystem Species and Community Profiles: Life Histories and Environmental Requirements of Key Plants, Fish and Wildlife.* Prepared by the San Francisco Bay Area Wetlands Ecosystem Goals Project. (ed Olofson PR). San Francisco Bay Regional Water Quality Control Board, Oakland, Calif., pp. 184–192.

Margulis, L (1971) Whittaker's five kingdoms of organisms: minor revisions suggested by considerations of the origin of mitosis. *Evolution* 25(1): 242–245.

Martin G (2011) The Middle Way. *Bay Nature*, July-September 2011, https://baynature.org/article/the-middle-way/; accessed September 27, 2020.

Martín-Durán JM, Passamaneck YJ, Martindale MQ, Hejnol A (2016) The developmental basis for the recurrent evolution of deuterostomy and protostomy. *Nature Ecology & Evolution* 1: 0005.

Mathiasson ME, Rehan SM (2019) Status changes in the wild bees of Northeastern North America over 125 years, https://doi.org/10.1111/ icad.12347.

Mayer JJ, Brisbin IL Jr (2008) *Wild Pigs in the United States: Their History, Comparative Morphology, and Current Status.* University of Georgia Press, Athens, Georgia, p. 20.

McKercher L, Gregoire DR (2020) *Lithobates catesbeianus* (Shaw, 1802): U.S. Geological Survey, Nonindigenous Aquatic Species Database, Gainesville, Florida.

Melquist WE, Dronkert AE (1987) River otter. *Wild Furbearer Management and Conservation in North America* (eds. M Novak, Baker JA, Obbard ME, Malloch B) Ontario Ministry of Natural Resources, Toronto, Canada, pp. 626–641.

Miller CR, Waits LP (2006) Phylogeography and mitochondrial diversity of extirpated brown bear (*Ursus arctos*) populations in the contiguous United States and Mexico. *Molecular Ecology* 15: 4477–4485.

Moffett MW (2012) Supercolonies of billions in an invasive ant: What is a society? *Behavioral Ecology* 23: 925–933.

Moritz C, Schneider CJ, Wake DB (1992) Evolutionary relationships within the *Ensatina eschscholtzii* complex confirm the ring species interpretation. *Systematic Biology* 41: 273–291.

Morris RH, Abbott DP, Haderie EC (1980) *Intertidal Invertebrates of California.* Stanford University Press, Palo Alto, California.

Moyle PB, Israel JA, and Purdy SE (2008) *Salmon, Steelhead, and Trout in California: Status of an Emblematic Fauna, Center for Watershed Sciences.* University of California Davis, Davis, California, pp. 89–91.

Mulvaney D (2013) *Green Atlas: A Multimedia Reference.* SAGE Publications, Thousand Oaks, California, p. 32.

Murray W (2004). *Elsevier's Dictionary of Reptiles.* Elsevier, Amsterdam, The Netherlands, p. 122.

National Ocean Service (2013) What lives in a kelp forest: Kelp forests provide habitat for a variety of invertebrates, fish, marine mammals, and birds. NOAA. Updated January 11, 2013. accessed September 27, 2020.

National Ocean Service (2020) Kelp forests: A description. https://sanctuaries.noaa.gov/visit/ecosystems/kelpdesc.html; accessed September 27, 2020.

Neilson S (2019) More bad buzz for bees: Record number of honeybee colonies died last winter. KQED. Retrieved from: https://www.npr.org/sections/thesalt/2019/06/19/733761393/more-bad-buzz-for-bees-record-numbers-of-honey-bee-colonies-died-last-winter#:~:text=Bee%20decline%20has%20many%20causes,systems%20and%20can%20kill%20them.

NOAA (2011) National Oceanic and Atmospheric Agency: San Diego climate by month, U.S. Department of Commerce National Oceanic & Atmospheric Administration National Environmental Satellite, Data, and Information Service. https://www.wrh.noaa.gov/sgx/climate/san-san-month.htm.

Norwak RM (1999) *Walker's Mammals of the World.* Johns Hopkins University Press: Charles Village, Baltimore, MD, p. 1521.

NPS (2018) Marine invertebrates. National Park Service. https://www.nps.gov/goga/learn/nature/marine-invertebrates.htm.

Oksanen L, Fretwell SD, Arruda J, Niemala P (1981) Exploitation ecosystems in gradients of primary productivity. *American Naturalist* 118: 240–261.

Ordeñana MA, Crooks KR, Boydton EE, Fisher RN, Lyren LM, Siudyla S, Haas CD, Harris S, Hathaway SA, Tureschak GM, Miles AK, Van Vuren DH (2010) Effects of urbanization on carnivore species distribution and richness. *Journal of Mammalogy* 91(6): 1322–1331.

Ort BS, Cohen S, Boyer KE, Wyllie-Echeverria S (2012) Population structure and genetic diversity among eelgrass (*Zosteria marina*) beds and depths in San Francisco Bay. *Journal of Heredity* 103: 533–546.

Ortiz JL, Muchlinski AE (2014) Urban/suburban habitat use by a native and invasive tree squirrel. *Bulletin of the Southern California Academy of Sciences* 113: 116.

Ostrom JH (1973) The ancestry of birds. *Nature* 242(5393): 136.

Padian K (1986) The origin of birds and the evolution of flight. *Memoirs of the California Academy of Sciences* 8: 1–55, California Academy of Sciences, San Francisco, California.

Parr I (2019) Naturally, 2019 Closes with Thousands of 10-Inch Pulsing "Penis Fish" Stranded on a California Beach. *Bay Nature.* Retrieved from: https://baynature.org/2019/12/10/naturally-2019-closes-with-thousands-of-10-inch-pulsing-penis-fish-stranded-on-a-california-beach/.

Paton TA, Baker AJ, Groth JG, Barrowclough GF (2003) RAG-1 sequences resolve phylogenetic relationships within charadriiform birds. *Molecular Phylogenetics and Evolution* 29: 268–278.

Pecon-Slattery J, O'Brien SJ (1998) Patterns of Y and X chromosome DNA sequence divergence during the Felidae radiation. *Genetics* 148: 1245–1255.

Pereir, RL, Wake, DB (2009) Genetic leakage after adaptive and nonadaptive divergence in the *Ensatina eschscholtzii* ring species. *Evolution* 63(9): 2288–2301.

Pons J-M, Hassanin A, Crochet P-A (2005) Phylogenetic relationships within the Laridae (Charadriiformes: Aves) inferred from mitochondrial markers. *Molecular Phylogenetics and Evolution* 37: 686–699.

Poppe KL, Rybcyk JM (2018) Carbon sequestration in a Pacific Northwest eelgrass (*Zostera marina*) meadow. *Northwest Science* 92(2): 80–91.

Poulter G (2010) *Becoming Native in a Foreign Land: Sport, Visual Culture, and Identity in Montreal, 1840–1885.* UBC Press, University of British Columbia, Vancouver, British Columbia, p. 33.

Raftery M (2013) Are there bears in the woods? In San Diego and East County, the answer may be "yes", East County Magazine April 2013, https://www.eastcountymagazine.org/are-there-bears-woods-san-diego-and-east-county-answer-may-be-yes; accessed February 22, 2021.

Rebman JP, Simpson MG (2014) *Checklist of the Vascular Plants of San Diego County*. 5th edition. San Diego Natural History Museum and San Diego State University, http://www.sci.sdsu.edu/plants/simpson/pdfs/Rebman_Simpson2014-ChecklistPlantsSan-DiegoCo.pdf; accessed March 8, 2021.

Resh VH, Cardé RT (2009) *Encyclopedia of Insects*. Academic Press, Cambridge, Massachusetts, p. 722.

Reynolds JW (2016) Earthworms (Oligochaeta: Acanthodrilidae, Lumbricidae, Megascolecidae, Ocnerodrilidae and Sparganophilidae) in the Central California Foothills and Coastal Mountains Ecoregion (6), USA. *Megadrilogica* 21: 73–78.

Rice DW (1998) *Marine Mammals of the World. Systematics and Distribution*. Special Publication Number 4. The Society for Marine Mammalogy, Lawrence, Kansas

Ripple WJ, Estes JA, Beschta RL, Wilmers CC, Ritchie EG, Hebblewhite M, Berger J, Elmhagen B, Letnic M, Nelson MP, Schmitz OJ, Smith DW, Wallach AD, Wirsing AJ (2014) Status and ecological effects of the world's largest carnivores. *Science* 343: 1241484.

Robbins CS, Brunn B, Zim HS (1983) *Birds of North America; A Guide to Field Identification*. Golden Press, New York, NY.

Roth B, Sadeghian P (2006) *Checklist of the Land Snails and Slugs of California*. 2nd Edition. Santa Barbara Museum of Natural History, Santa Barbara, CA.

Rubenstein S (2020) More than a dozen California condors missing after wildfire destroys their Big Sur sanctuary. *San Francisco Chronicle*, August 26, 2020; https://www.sfchronicle.com/california-wildfires/article/Fires-destroy-Big-Sur-condor-sanctuary-15516997.php.

Rychel AL, Smith SE, Shimamoto HT, and Swalla HT (2006) Evolution and development of the Chordates: collagen and pharyngeal cartilage. *Mollecular Biology and Evolution* 23(3): 541–549.

Sahagún L (2020) Southern California mountain lions get temporary endangered species status. *Los Angeles Times*. https://www.latimes.com/environment/story/2020-04-16/state-panel-studying-threatened-species-protection-for-southern-california-cougars.

San Diego Zoo News report (ND) California Condor – San Diego Zoo Animals & Plants. https://animals.sandiegozoo.org; accessed March 30, 2020.

Santos RL (1997) *The Eucalyptus of California*. Alley-Cass, Denair, California.

SDMMP (2010) San Diego Management & Monitoring Program: Mountain Lion, San Diego Management & Monitoring Program, San Diego, California, https://sdmmp.com/species_profile.php?taxaid=552479; accessed March 7, 2021.

SDPARKS (ND) Invasive Species, County of San Diego Parks and Recreation, San Diego, California, https://www.sdparks.org/content/sdparks/en/news-events/news-archives/InvasiveSpecies.html; accessed March 14, 2021.

SelecTree (2020) SelecTree. California buckeye. *Aesculus californica* Tree Record. 1995–2020, https://selectree.calpoly.edu/tree-detail/aesculus-californica.

Shelley RM (2002) Annotated checklist of the millipedes of California (Arthropoda: Diplopoda). *Monographs of the Western North American Naturalist* 1: 90–115.

Sibley DA (2000) *The Sibley Guide to Birds*. Alfred A. Knopf, New York, NY.

Sillero-Zubiri C, Hoffman Michael, MacDonald DW (2004) *Canids: Foxes, Wolves, Jackals, and Dogs: Status Survey and Conservation Action Plan*. IUCN, Gland, Switzerland and Cambridge, UK, p. 95.

Simmons E (2019) The importance of having insects. Bay Nature May 2019.

Smith CC, Haglund TR, Ruiz M, Fisher RN (1993) The status and distribution of freshwater fishes of southern California, Bull. Southern California Acad. Sci. 92(3): 101–167.

Somma, LA (2021) Pseudacris sierra (James, Mackey, and Richmond, 1966), *US Geological Survey, Nonindigenous Aquatic Species Database*. Gainesville, Florida, https://nas.er.usgs.gov/queries/FactSheet.aspx?SpeciesID=2780. accessed March 13, 2021.

South A (1992) *Terrestrial Slugs: Biology, Ecology and Control.* Chapman and Hall, London, UK, pp. 1–428.

Stacey BJ (ND) Birds of San Diego County, California, iNaturalist Guides, www.inaturalist.org/guides/408; accessed March 7, 2021.

Stebbins RC (1949) Speciation in salamanders of the plethodontid genus *Ensatina. University of California Publications in Zoology* 48: 377–526.

Stebbins RC (1954) Natural history of the salamanders of the plethodontid genus *Ensatina. University of California Publications in Zoology* 54: 47–124.

Stebbins RC (1959) *Reptiles and Amphibians of the San Francisco Bay Region.* University of California Press, Berkeley and Los Angeles.

Stebbins RC (2003) *A Field Guide to Western Reptiles and Amphibians*, 3rd Edition. The Peterson Field Guide Series. Houghton Mifflin Company, Boston and New York

Stebbins TD, Eernisse DJ (2009) Chitons (Mollusca: Polyplacophora) known from benthic monitoring programs in the Southern California Bight. *The Festivus*, (Special Issue) 41(6): 53–100 (with errata).

Stebbins RC, McGinis SM (2012) *Field Guide to Amphibians and Reptiles of California: Revised Edition (California Natural History Guides).* University of California Press, Berkeley, California.

Sunset (1995) *Sunset Western Garden Book; 40th Anniversary Edition. June 1995*. Sunset Publishing Corporation, Menlo Park, CA 94025; pp. 28–31.

Tan A-M, Wake DB (1995) MtDNA phylogeography of the California newt, Taricha torosa (Caudata, Salamandridae). *Molecular Phylogenetics and Evolution* 4: 383–394.

Tighe D (2019) Meet the Bay's incredible swimming worms. *Bay Nature*. https://baynature.org/2019/08/06/meet-the-bays-incredible-swimming-worms/.

UCANR (2021) *California Snails and Slugs, Division of Agriculture and Natural Resources*. University of California, Riverside, California, https://ucanr.edu/sites/CalSnailsandSlugs/; accessed March 13, 2021.

UCR News (2015) The Top Nine Scariest Southern California Bugs, UC Riverside News, University of California Riverside, Riverside, California, https://news.ucr.edu/articles/2015/10/20/top-nine-scariest-southern-californian-bugs; accessed March 13, 2021.

Ueda KI, Agarwal RG (ND) California sea slugs - Nudibranchs (and other marine Heterobranchia) of California. *iNaturalist*. https://www.inaturalist.org/guides/40.

van Tuinen, M, Waterhouse DM, Dyke GJ (2004) Avian molecular systematics on the rebound: a fresh look at modern shorebird phylogenetic relationships. *Journal of Avian Biology* 35: 191–194.

Wake DB (1997) Incipient species formation in salamanders of the Ensatina complex. *Proceedings of the National Academy of Sciences* 94: 7761–7767.

Wallach AD, Johnson CN, Ritchie EG, O'Neill AJ (2010) Predator control promotes invasive dominated ecological states. *Ecology Letters* 13: 1008–1018.

Watanabe J (2017) Molluscs, snails, nudibranchs, bivalves, chitons, Nearshore Plants and Animals of the Monterey Bay, Seanet, Hopkins Marine Station, Pacific Grove, California, https://seanet.stanford.edu/Molluscs; accessed February 26, 2021.

Watkins B (ND) California marine flatworms. *California Diving*. https://cadivingnews.com/california-marine-flatworms/.

Wayne RK, Geffen E, Girman DJ, Koepfli KP, Lau LM, Marshall CR (1997) Molecular systematics of the Canidae. *Systematic Biology* 46(4): 622–653.

WERC (ND) Western Ecological Research Center, American Badgers in San Diego County, USGS, US Dept. Interior, Washington, DC, https://www.usgs.gov/centers/werc/science/american-badgers-san-diego-county?qt-science_center_objects=0#qt-science_center_objects; accessed March 7, 2021.

Whittaker RH (1969) New concepts of kingdoms or organisms. *Science* 163(3863): 150–160.

Williams BL, Brodie ED III (2003) Coevolution deadly toxins and predator resistance: self-assessment of resistance by garter snakes leads to behavioral rejection of toxic newt prey. *Herpetologica* 59: 155–163.

Wilson DE, Mittermeier RA, eds. (2009). *Handbook of the Mammals of the World*, Volume 1: Lynx Ediciones, Carnivora, Barcelona, pp. 50–658.

Wolf KM (2015) Management of blue gum eucalyptus in California requires region-specific consideration. *California Agriculture* 70: 39–47.

Woodburne MO (2004) *Late Cretaceous and Cenozoic mammals of North America: Biostratigraphy and Geochronology*, (ed Woodburne MO), Columbia University Press, New York, NY.

Woodward SL, Quinn JA (2011) . *Encyclopedia of Invasive Species: From Africanized Honey Bees to Zebra Mussels*, ABC-CLIO: Santa Barbara, California; Denver, Colorado, Oxford, UK.

Wozencraft WC (2005) Order Carnivora. In: *Mammal Species of the World: A Taxonomic and Geographic Reference,* 3rd Edition (eds. Wilson DE, Reeder DM) Johns Hopkins University Press, Baltimore, Maryland, pp. 624–628.

9 Restoring the Bay

THE BAY IS NOT WHAT IT ONCE WAS

San Diego Bay was not always there. The current Bay took millions of years to make. By about 2 million years ago, the movement of the crustal plates had created the basic physical structure that surrounds the Bay. Since then, other forces have refined and modified that structure. The water in the Bay has disappeared and reappeared several times as sea levels rise and fall. The most recent flooding of the estuary occurred 10–11,000 years ago.

Then humans arrived. More importantly, in the last 150 years, Europeans and others arrived, and they have caused the Bay to deteriorate significantly. They could not change the basic structure, but they did change the outline significantly. Large amounts of the tidal marsh have been lost to filling and building. Areas were filled for development for homes, business, industry, and the military. Islands were connected. Bays dredged and filled. Rivers and streams and even the Bay have been filled with silt and pollution. More modern efforts have attempted to contain rivers within their beds to reduce the threats of flooding. Pollutants have been swept into the Bay from sewage systems (or simple raw sewage), industrial processes, and agricultural activities.

Now human activities are resulting in climate change, and the Earth is warming at an increasing rate. As the ice caps melt, sea levels rise, rain patterns change, the temperature of bodies of water are increasing, freshwater becomes less available, and changes occur to many other systems. All of this change will stress the plants and animals around the world and in the Bay. Global warming and rising ocean levels are likely to threaten even more wetlands. The Bay has become silting up by mining and other practices. Pollution comes from development, agriculture, and industrialization.

Fortunately, in the last few decades, efforts have begun to maintain and even restore natural areas, such as wetlands, marshes and salt, and mudflats. New laws, such as the Clean Water Act of 1972, California's Porter Cologne Act of 1969, the McAteer-Petris Act of 1965, and others, are helping. These new efforts focus on many aspects, such as controlling pollution and contamination and restoring the bay shore. But the work will not be easy. To show how hard this is, a study in the San Francisco Bay Area (Stralberg et al., 2011) concluded that bay lands and mid-marsh areas could be restored with a reasonable amount of material. However, significantly more material would be needed to save those same wetlands over 100 years.

There is plenty of work to do on the restorations. Salt ponds can be brought back to a more natural state. Wetlands can be restored along the shore. Rivers can

DOI: 10.1201/9780429487460-9

be made more healthy and less polluted. Military bases and industrial sites can be cleaned up. Restorations are important, but they require careful planning and execution, and real determination and political will to see them through. Finally, they need money. Fortunately, there are some hopeful signs as noted below.

RESTORATION

What does it actually mean to restore an ecosystem? What are the metrics that one would use? How can one measure the function of an ecosystem? In fact, direct tests are rare, and it isn't completely clear how to answer those questions. Although achieving functional equivalency of restored and natural ecosystems is a desirable restoration goal, direct assessments of function are rare, as are data supporting the use of indicators of function. The issue is how structural attributes can be used to assess ecosystem functioning. A functional equivalency index mixed both structural and functional measures, and an often-cited model of ecosystem degradation and restoration depicted a straight-line relationship between the two variables. Several points need clarification, not only for tidal wetlands but for restoration ecology generally.

Zedler and Lindig-Cisnerow (2000) state that it is not realistic to hope for a linear relationship among structure, function, and time and that one cannot hope that equivalent structure means equivalent function. The measures and relationships are far more complicated. For example, they suggest that measuring soil organic matter and total nitrogen levels may not be appropriate measures. They do suggest some helpful measures, including soil texture soil organic matter and nutrients, vegetation structure, invertebrate and fish populations, and topographic complexity.

Mossman et al. (2012) compared 18 deliberately realigned and 17 accidentally realigned sites with 34 natural reference saltmarshes in the United Kingdom. They found that halophytic species adapted rapidly to the managed marshes, but the composition of the communities was still different than the natural communities. The oxygen and humidity levels might have been lower in the managed marshes so that they were less welcoming to diverse species. In contrast, the accidentally realigned marshes were quite similar to the natural marshes. They conclude that it might be unrealistic to expect that managed marshes will meet the current standards and that changes to those standards might be warranted.

Restoration is important, but it is also a great strategy to maintain wild areas. San Diego County is home to 200 plants and animals at risk, including the least Bell's vireo and California gnatcatcher, the arroyo southwestern toad, Stephens' kangaroo rats, and San Diego fairy shrimp. Working with public and private partners, the Nature Conservancy has had great success in preserving wildlands around San Diego for the benefit of those and more species. The San Diego National Wildlife Refuge has been expanded. The San Diego Backcountry comprises rolling hills, grasslands, mountains, and deserts. The Nature Conservancy acquired numerous plots, including the 5,400-acre Santa Ysabel Open Space Preserve, the 4,500-acre Cañada de San Vicente Preserve and the 4,000-acre

Ramona Grasslands preserve. They also helped to expand and connect key areas, such as Anza-Borrego Desert State Park, Cuyamaca Rancho State Park, and Volcan Mountain. Near the Mexico border, they have been arranging wildlife corridors to help species that need large ranges, such as mountain lions, to move easily between the two countries.

Complexity of Restoration

It is easy to talk about restoring wetlands and other areas, but it is not easy to do that. Wetlands are very complex systems that have developed over millions of years. They cannot be recreated by simply mixing back some of the components. In most cases, the complete census of organisms is not even known. The loss of environments and other challenges are complicated further by the isolation of the relatively small pockets of habitat that remain. Communication between those pockets limits cross-fertilization and future diversity.

A number of strategies have been developed to overcome these problems and restore habitats. Moreno-Mateos et al. (2020) suggest that the focus must be on very long-term (hundreds to thousands of years) programs to reestablish the habitat. An effort on this time scale gives an opportunity to adjust more effectively. They also suggest a whole-genome sequencing approach to determine what functions are needed in the habitat. This intriguing strategy is probably the best way to proceed, but clearly, it would require extraordinary patience and funding to see through.

An obvious question is how do we measure progress in reestablishing an environment. Ruiz-Jaen and Aide (2005) discuss this problem. They point out that most studies evaluate diversity, vegetation structure, and ecological processes. They suggest adding variables within each of these three measures that clearly related to ecosystem functioning and establishing two reference sites to understanding the variation in ecosystems.

Wetlands

Around San Diego Bay, 75–90% of the wetlands have been lost. Most of the communities around the Bay have only small amounts of wetlands. Still, any success in preserving or reestablishing wetlands is a big win for the environment.

The restoration of wetlands has many challenges. Development often fragments natural habitats so that they become isolated. They may be cut off from more established wetlands, and so, they will be less likely to be repopulated by native species from the established wetlands. Biodiversity is thus limited. The loss of diversity might lead to a less complex canopy, more opportunities for invasive species to take root, and less nitrogen accumulation. Small urban salt marshes are highly vulnerable to exotic species. For example, some time back, a researcher imported white mangrove (*Avicennia marina*) into Mission Bay Marsh to study it. However, the plants grow to 2 meters tall and would attract raptors that might prey on the chicks of the endangered clapper rail. Fortunately, scientists were able to remove all of the mangrove before it took firm hold, but it was a labor-intensive and long project. International ports also attract lots of travelers and traders, and some may bring exotic species with them. Development also disrupts the flow of water and sediment. That changes the characteristics of the wetlands so that some species might be lost. Although they are not always as noticeable as wetlands, the transition zones that exist between the wetlands and the terrestrial areas are also very important to the health of the wetlands. Some species need both habitats.

Morzaria-Luna and Zedler (2007) studied how the ways that seeds are dispersed affects the restoration of salt marshes. Their work focused on an 8-hectare salt marsh being restored and another established marsh, both in the Tijuana Estuary just south of San Diego. Seed dispersal was more effective in the tidal areas than the nontidal areas. Specifically, the dominant species, *Sarcocornia pacifica*, was also the species with the most seedlings. The restored areas had many fewer seedlings, and many of those seedlings were invasive species. Replanting would be more successful if it were done in the winter when the native species enjoyed the maximum tidal dispersal.

The San Diego Bay National Wildlife Refuge has been collaborating with San Diego Gas and Electric Company to restore coastal wetland habitat at the D Street Fill. They hope to benefit native fish, wildlife, and plant species and provide habitat for migratory shorebirds and salt marsh-dependent species. The project involves 11 acres.

Dredging

San Diego Bay is occasionally dredged to clear the shipping channels. Other regions, such as Mission Bay, have been the subject of extensive dredging. Dredging is important, but it can also be a serious problem to wildlife, and there is a question about how to deal with the spoils.

Dredging invariably stirs up a cloud of sediment that can drift in all directions, as determined by the winds and waves, turbidity, characteristics of the sediments, and more (Capello et al., 2010). By better understanding how sediments move, researchers can help others to minimize the detrimental effects of dredging on marine life (Fraser et al., 2017). Dredging increases the amount of suspended sediment and reduces light levels underwater. This affects marine organisms that depend on photosynthesis. The additional suspended material interferes with feeding and other activities. Oysters, for example, are very sensitive to dredging (Wilber and Clarke, 2010). They can be physically covered by the settling material. The sediment can interfere with the oyster's filter feeding. Finally, oyster larvae need clean hard surfaces to attach to. Even small amounts of loose sediment can disrupt the attachment and damage the oyster population. In most areas, new regulations limit the amount of dredging around oysters.

For many years, dredged material was considered waste to be disposed of. However, in recent years, sediment has come to be seen as a precious resource that is essential for keeping the estuary healthy (Carse and Lewis, 2020). It can be used to restore shorelines, beaches, and wetlands. In addition, sediment is continually washed out to sea. New sediment, either from natural sources or dredging, can help to maintain a healthy balance of material lost and gained (Milligan and Holmes, 2017). Unfortunately, a loss of that balance has large-scale effects, and human interventions have often been the reason for the disruption. Dams and other structures upstream interrupt the flow of water and sediment. Dredging can also remove sediment or redistribute it to areas that need more.

One method of preserving beaches is to add sand to them. Ludka et al. (2018) looked at four Southern California beaches that received new sand offshore. The sand was several meters thick, and the pad was 500–1500 meters long. Three of these pads contained coarse sand. After several years of severe winter storms, the sand had formed accretion crowns just offshore that protected the beach. A fourth beach, the sand protected the beach by collecting on the riprap on the beach. The sand remained in position for varying times, but in all cases, it would have to be replaced.

Contamination of sediment is another serious issue. Many pollutants bind easily to sediments and accumulate at the bottom of the Bay. One common contaminant is polychlorinated biphenyls (PCBs) (Yee and Wong, 2019). PCBs reach the Bay in run-off from storm drains mainly. Importantly, PCBs and other contaminants can be concentrated in fish and seafood and make them unsafe for human consumption. Recent efforts to reduce contaminants have helped to greatly reduce PCB levels in the Bay. Also, sediment that is to be used to restore wetlands must be tested for levels of contaminants to make sure it is safe to distribute.

Pollution can come from surprising sources. Recreational boating is very popular on San Diego Bay. The antifouling coatings that are applied to the hulls of the boats contain copper to retard the growth of algae, barnacles, and corals. Unfortunately, copper is also harmful to marine life, and the copper leaches from the boat hulls over time. Carson et al. (2009) explored the issues involved in producing a policy to phase out those coatings. The alternatives include banning them or providing economic incentives to get rid of them. Ciriminna et al. (2015) describe advances in nanochemistry that are providing much more eco-friendly antifouling coatings than copper-based ones.

Wildlife

Wild areas often have two mandates. They are meant to conserve the plants and animals in that area, and they are to provide recreation areas for people. Unfortunately, these two requirements are not always in synch. In particular, even careful human activities can be detrimental to native species. Human activities in natural areas stress nature animals. Even non-motorized activities can disrupt normal routines, affect feeding, and displace animals from their habitat. Some animals attempt to avoid humans, such as bobcat (*Lynx rufus*), grey fox (*Urocyon cinereoargenteus*), and mountain lion (*Puma concolor*). Others adapt to the presence and even thrive, such as coyote (*Canis latrans*) and raccoon

(*Procyon lotor*). The results of these studies can help recreation park managers to limit the impact on wildlife while still allowing people the opportunity to experience wild areas. San Francisco is an excellent location to study these interactions. Reilly et al. (2017) also looked at how hiking, cycling, horse riding, and dog walking affected 10 species of mammals in the San Francisco Bay Area. They set camera traps at 241 locations in 87 protected areas to compare human and mammal activity in areas that were or were not used by humans and during day and night. When dogs were present, mountain lions and opossums were seen less frequently. Coyotes visited high-use areas at night, but not in the day. Small carnivores that hunt at night were not affected by daylight activities. They suggested prohibiting dogs and establishing buffers between areas used by humans and those by animals.

To assess the extent of this problem in San Diego, Reed et al. (2014) examined how human activities affected sensitive species in the reserves there. Most interestingly, they sought to document those effects in an experiment in which 92 specific sites were examined before and after human activities. They observed that a decline in reptile species richness was associated with human activities. Smaller bodied lizards (e.g., common side-blotched lizard, *Uta stansburiana*) were less common in areas with high levels of human However, the western fence lizard (*Sceloporus occidentalis*) was not affected. Human activity was also associated with less activity by bobcats and mule deer. Furthermore, bobcats, gray foxes, mule deer, and raccoons were less active in areas frequented by humans. With additional population growth and higher expected rates of human use of the parks, the effects on wildlife are expected to be even greater in the future. Many residents of San Diego are not aware of the effects of humans on wildlife. In 2018, a survey was conducted as part of the San Diego End Extinction initiative. It found that 71% of the 600 people surveyed knew little about the threat to plants and animals in the region (Tinkler et al., 2019).

As our urban areas expand further into previously wild habitats, the pressure on wild animals increases. Human activities in natural areas stress nature animals. Even non-motorized activities can disrupt normal routines, affect feeding, and displace animals from their habitat. Most wild animals typically avoid urbanized areas, but some have adapted in amazing ways. Others adapt to the presence and even thrive, such as coyote (*Canis latrans*) and raccoon (*Procyon lotor*).

Mammalian carnivores are particular sensitive to habitat loss and fragmentation. They typically need large ranges and low densities to find sufficient prey. Over the last 150 years, the San Diego region has developed into one of the largest metropolitan areas in the United States. Much of the traditional habitat for animals has been lost, and what remains can be separated by development and highways. Wild animals try to adapt by various means. Some come to depend on food left by humans. That could include trash, fruit trees, and even domesticated pets. Amazingly, a number of large carnivores have managed to live in the region, including bobcats (*Lynx rufus*), coyotes (*Canis latrans*), and gray foxes (*Urocyon cinereoargenteus*). Coyotes may be the most successful of the group. That might be because they are less particular about their diet and will eat almost anything.

Larson et al. (2015) studied the scats of bobcats, coyotes, and gray foxes in the urban areas of Southern California to determine their diet habits. Coyotes ate a broad range of foods, including mammals, fruits, seeds, birds, and invertebrates. Cats were a favorite item. Outdoor cats are a serious threat to birds, and ironically, the coyote predation might help with maintaining the bird population. Foxes had a diet similar to the coyotes, except that they rarely take cats. They do very well in urban areas, especially those with a low number of competing coyotes. Bobcats are more picky in their food. They are essentially carnivores. They take mostly rodents and rabbits, but will also have invertebrates and birds. They rarely invaded trash cans or attacked cats. Clearly, their more selective diet limits their ability to live in urban areas. Like all science, identification of scats is not easy. Morin et al. (2016) showed that it is easy to misidentify scats by morphology alone. They suggested that, if possible, future studies should incorporate some form of noninvasive genetic testing to ensure correct identification.

One challenge to restoring ground-nesting waterbird populations has been the removal of predators. Humans introduced some of those predators, including feral cats (*Felis domesticus*) and the Norway rat (*Rattus norvegicus*). Others are native, such as the striped skunk (*Mephitis mephitis*). Meckstroth and Miles (2005) examined strategies in the San Francisco Bay. They looked at the number of nests and the hatching success in sites where predators were removed and sites where they were not removed. Nests in the removal sites had greater nest densities, but lower hatching success. They had some success with feral cats, but striped skunks could not be controlled. predator composition and abundance remained the same, except for feral cats. The researchers concluded that there were so many skunks that they could easily repopulated the sites. The California least tern is nests on coastal estuaries and beaches of San Diego and California. They are prey for multiple native and nonnative animals, and various programs have attempted to remove or control the predators of these endangered birds (SDBNWR, nd). The most common are feral cats, striped skunks, Virginia opossums, California ground squirrels, common ravens, western gulls, American kestrels, and ham owls (Carrillo, 2004).

SPECIFIC RESTORATION PROJECTS

South San Diego Bay Coastal Wetland Restoration and Enhancement Project

The Project restored and improved 261 acres of coastal wetland habitat at the south end of San Diego Bay. Before restoration, the water quality was poor, and the two salt ponds had low diversity and abundance of fish and other marine species. The levees of the Western Salt Ponds were breached to connect the area with the Otay River channel and the Bay, and the elevations for the habitat were achieved (Figure 9.1). The result is an estuary with open-water, intertidal and upland marsh habitats. Native salt marsh vegetation was planted.

FIGURE 9.1 Restoring the Salt Ponds. Otay River water flows into the newly breached former salt ponds on the San Diego Bay NWR. The main channel of water here is a healthy mix of bay water and freshwater from the Otay River, which hasn't been able to flush the western salt ponds since the 1950s (Photograph courtesy of Lisa Cox and the US Fish and Wildlife Service).

The monitoring results at the Western Salt Ponds are encouraging. The tidal amplitude and water quality in the ponds are now similar to those in south San Diego Bay the number of species of fish, plants, birds, and marine mammals are increasing each year. The restored ponds are an excellent nursery for fish and invertebrates. Cordgrass (*Spartina* sp.), eelgrass (*Zostera* sp.)—a favorite food for green sea turtles (*Chelonia mydas*)—and other native vegetation are taking hold in a natural manner. Pacific pickleweed (*Salicornia pacifica*) and Bigelow's pickleweed (*Salicornia bigelovii*) are doing well. Ridgway's rails (*Rallus obsoletus*) are breeding in the Otay River channel. Round stingray (*Urobatis helleri*), California halibut (*Paralichthys californicus*) and slough anchovy (*Anchoa delicatissima*) are typical of the Bay. Polycheates, oligochetes, and crustaceans include the California horn snail (*Cerithideopsis californica*), California jackknife clam (*Tagelus californianus*), and Asian mussels (*Perna viridis*, a nonnative).

The efforts to improve the Chula Vista Wildlife Reserve also achieved success. Excavation of a channel is expected to improve water quality with a greater tidal influence over time. The goal for cover by vegetation is to be similar to the Tijuana Estuary. Cordgrass and Bigelow's pickleweed have been migrating to the site. Fish and invertebrate species are similar to other Southern California bays and lagoons.

Sweetwater Marsh Restoration

Sweetwater Marsh is a 316-acre parcel that includes habitat for several endangered and threatened species: California least tern, Ridgway's rail, western snowy plover, and Belding's savannah sparrow, and an endangered plant salt marsh bird's beak. The area contains a deserted rail line and rail support facility, slaughterhouses, manufacturing, and a burn dump. Earlier remediations attempted to clean up contaminated soil. Chemicals of concern included metals, dioxins and furans, pesticides, polychlorinated biphenyls, polycyclic aromatic hydrocarbons, and lead. This latest clean-up is just beginning (Figure 9.2).

Gunpowder Point section of Sweetwater Marsh is an interesting case. From 1916–1920, the Hercules Powder Company used the site to produce potash and acetone by fermenting kelp. Potash is an ingredient of black powder, and acetone is used to extrude cordite, a smokeless gunpowder. The site had 156 redwood fermentation tanks (190,000 liters each), nine storage tanks (1,500,000 liters each), and underground concrete vaults (millions of liters). There was also an onsite railroad system, pipelines, a laboratory, and other facilities.

FIGURE 9.2 Sweetwater marsh. This marsh represents a successful restoration of wetlands in the San Diego Bay region (Photograph courtesy of the US Geological Survey).

Otay River Restoration

This restoration is just beginning and will have many benefits. The *Otay River Estuary Restoration Project* will restore 125 acres of coastal wetland and upland habitats at two locations to benefit native fish, migratory birds, and other coastal-dependent species. The *River Partners Project*is a collaboration by River Partners, US FWS, and San Diego National Wildlife Refuge Complex to replace non-native vegetation with diverse, native plants to support a variety of migratory birds and other native wildlife. It will also provide a natural filter for stormwater runoff.

MAJOR CHALLENGES

Invasive Species

The California Department of Fish and Wildlife (CDFW, 2021) lists a number of marine invasive species. Invasive species arrive by several routes. Some attach directly to ship hulls and drop off or release larvae in distant harbors. Others travel in the ballast water loaded on-board ships to adjust buoyancy and provide stability. The ballast is then dropped at the port of call and allows those organisms to become established in a non-native region. Still, others are brought intentionally for new populations for fisheries or aquaculture, but the new species is trouble for the new ecosystem. Some organisms actually live on the shells of oysters. When oysters were brought to a new site, the hitchers are inadvertently introduced.

Some are listed here. The green crab *(Carcinus maenas)* is native to Europe, but has made it to America. It preys on native clams, oysters, mussels, and crabs. Asian kelp *(Undaria pinnatifida)* is from Japan, China, and Korea and arrived by hull fouling. It outcompetes native species. The Chinese mitten crab *(Eriocheir sinensis)* is from China and Korea and probably arrived in ballast water or by intentional releases. It interferes with fish passage and salvage, water treatment and power plants, and more. The clubbed tunicate (*Styela clava*) is from Eastern Asia and likely arrived on the hulls of ships or with imported oysters. This hardy crab can withstand salinity and temperature fluctuations.

Sea Level Rise

Sea level rise due to global warming is a serious threat to nearly all coastal areas. Estimates were that sea levels will rise 1.7 mm/year in the late 20th century to 3.1 mm/year in the early 21st century. Many low-lying areas will be at risk for inundation. Other aspects of climate change (e.g., storm intensity, rainfall, storm surges) will be exacerbated by climate change.

Scientists have been modeling these changes for some time. Different locations will be affected to greater or lesser degrees, and knowing what is likely to happen is critical for planning how best to adapt to those changes. The San Diego

Bay region is at great risk for inundation as sea levels rise. The average sea level has risen, and extreme tides are far more common. Scientists predict that sea levels are likely to rise more rapidly than previously thought, and the water supply will diminish with wetter winters and drier summers. These factors will make is more difficult to maintain shorelines and native species in the future.

Wetlands restoration projects add to the defense of the coast by providing additional areas for increased water levels to be absorbed even before they make contact with the seawalls, levees, and dikes that protect infrastructure and other items.

Local land subsidence may also make things worse, but maps displaying the risk of subsidence are lacking. Those correlated with areas of artificial landfill or Holocene mud deposits. They predicted an area of 125–429 km^2 is at risk of flooding, which is somewhat higher than when only sea level rise is considered. This more accurate mapping will help governmental agencies with planning for the future. Subsidence can be caused by a number of factors, including not watering soil, compaction, pumping out groundwater, and removing crude oil (Brandt et al., 2020). The use of metal-based coagulants to bind material together is one possible solution. To test this idea, one study (Stumpner et al., 2018) flooded wetlands with water containing coagulants that would accrete minerals. After 23 months of treatment, sediment samples were examined. Wetlands treated with polyaluminum chloride worked the best, but iron sulfate was also good. Other characteristics of the accreted material were also encouraging. Overall, these coagulants seem to have efficacy in reversing subsidence and storing carbon. By using them, levee failure might be reduced.

Improving the Air Quality

Unfortunately, San Diego has poor air quality. The city was ranked fifth for air quality by the Environmental Protection Agency in 2019 (EPA, 2019). Some improvement has been noted (USD, nd). The combined number of unhealthy air days and those for sensitive groups decreased from 25 in 2018 to 25 in 2019. The number of children hospitalized for asthma also decreased.

Auffhammer and Kellogg (2009) studied the effects of gasoline regulations on ozone levels in California. Those regulations add cost for customers and vary in their effectiveness in improving air quality. More specifically, they found that the federal regulations have no effect on emission of volatile organic compounds. However, the California regulations specifically require refiners to remove the chemicals that produce ozone, and thus, those regulations resulted in an improvement in air quality. The fuel used by ocean-going ships contains high levels of sulfur that yield air pollutants, such as sulfur dioxide and particulates. In early 2009, California began to require ships to use lower sulfur fuels.

One example of efforts to improve the air quality focused on the Barrio Logan, a section of the city adjacent to port facilities and that experiences heavy truck traffic (Karner et al., 2009). By concentrating on traffic operations, they were able to improve local air quality. They smoothed and rerouted truck traffic within the community. Emissions near the receptors were reduced, but the overall emissions were increased due to longer truck routes. This is an example of how cures sometimes cause other problems. In 2018, a community initiative, called Portside Environmental Justice Neighborhoods, began monitoring air quality.

Climate change is likely to cause the air quality to become worse in the near future (Steiner et al., 2013). Even now, excessive heat and drought have resulted in fires that made the air the worst in the world for a while in 2018. The continuing influx of people will also bring additional air pollution due to the continuing use of fossil fuels.

AREAS WITH SPECIFIC CHALLENGES

Tijuana River

Toxic flows of Mexican sewage have closed the beaches for kilometers, sickened residents and tourists, and melted the boots of border agents (Dooley, 2019). Yellow plumes flow out of the Tijuana River and drift north toward Imperial City and San Diego. According to a draft report from the Customs and Border Patrol, the sewage contains volatile organic compounds, uranium, iron, chromium, pesticides, herbicides, and biological contaminants.

It would be easy to blame the whole problem on Mexico, but it is much more complicated than that (Malone, 2020). From World War 2 to the 1960s, Tijuana's population exploded from 16,000 people to more than 340,000. Unfortunately, the sewage and stormwater handling capacities were not able to keep up with the growth. In addition, much of the growth of manufacturing around Tijuana is from investments from the United States and for industries that the US does not want. There was some recent good news. The new United States-Mexico-Canada Agreement will provide $300 million for infrastructure to stop the chronic flow of sewage across the border.

The Tijuana River is south of San Diego Bay, and it flows directly into the Pacific Ocean rather than the Bay. However, it is perhaps the most challenging area for restoration because the river is in two nations. It begins in Mexico, crosses the border, flows another 8 km, and empties into the Pacific Ocean. The final 3 km are a broad mud flat estuary called the Tijuana River Estuary that contains more than 370 species of birds. The river flows through the city of Tijuana, which is a heavily populated and industrialized section of Mexico. Essentially all of the water in the river is used in Mexico so that it only flows after heavy rains. Unfortunately, the heavy rains can also result in flooding and sewage runoff. The outflow of partially or untreated sewage is carried out into the Pacific, but it often causes beaches to be closed for kilometers. Sometimes the closures reach as far north as Coronado.

Mexico and the United States have been struggling with the problem of pollution on the River for decades. The challenges arise from the astounding population growth in Tijuana, different laws regarding pollution, different locations of the cause and effect, and the differences that occasionally arise between any two nations. Nevertheless, considerable cooperation has at least mitigated the pollution. In 1944, the two countries signed the United States-Mexico Water Treaty, which covered the Colorado and Tijuana Rivers and the Rio Grande. Mexico completed their flood control measures, but the US did not.

Despite many efforts to solve the problem, the river remains one of the most polluted. Raw and undertreated sewage is only one challenge. The water also contains DDT, hexavalent chromium, pathogens, and carcinogens. The Tijuana River is extremely polluted. The treatment infrastructure is simply inadequate to the challenge. Gersberg et al. (2004) characterized the quality of the water during wet and dry seasons. They used a test for toxicity that involved a water flea (*Ceriodaphnia dubia*). Under baseline conditions, the toxicity was low, but it increased after periods of rain. Their tests suggest that nonpolar organics are the source of the toxicity. They also speculated that the first hard rain washed out the toxicity into the river so that subsequent rain events produced less toxicity. Svejkovsky et al. (2010) used aerial imagery and other data to track stormwater runoff from the Tijuana River into the Pacific Ocean in 2003–2008. They could follow the runoff by its color on the surface of the ocean water. The discharge rate was the main determinant of fresh plumes. Bacterial levels were high in the fresh plumes. Oldest plumes contained both polluted and clean water. They concluded that high-resolution imagery is a valuable tool, but it cannot by itself identify polluted water plumes. Ayad et al. (2020) studied the discharge of stormwater and wastewater from the Tijuana River into the Pacific. They used remote imaging to differentiate ocean plumes of stormwater, wastewater, open ocean, and mixes of those. They showed that stormwater had the highest levels of dissolved organic matter, turbidity (12.4 to 45.7 FNU), and enterococcus bacteria. Microplastics are another serious pollutant that reaches the water through runoff. De Jesus Piñon-Colin et al. (2020) looked at microplastics at seven sites on the Tijuana River. The largest number were found near an industrial site, and they were mostly polymers. Rainfall increased the numbers.

The National Estuarine Research Reserve at Tijuana Estuary was established to prevent sediment from flowing into the Tijuana Estuary. Unfortunately, the sediment flows from Mexico, and the U.S. has no ability to limit the erosion in the watershed. The dominant species in the Estuary is pickleweed (*Salicornia virginica* L). However, a large amount of sediment arrives in the Estuary each year. Callaway and Zedler (2004) attempted to determine these effects. They sampled the sediment along the river and compared their results to previous surveys and weather records. The storms of 1994–95 were severe, and the amount of sediment deposited due to those storms was 10-fold greater than previous samples on the river and higher than samples collected elsewhere in Southern California. The high rates of accumulation have continued through 2003. Growth of the pickleweed has been hampered by the sediments. With the accumulation of sediment and the rise in elevation of the land, some of the wetlands have been converted to upland habitats with a loss of the pickleweed.

Mission Bay

Mission Bay is one of the most modified natural features in Southern California. It has been the subject of extensive dredging, filling, and development. It now features artificial shorelines, island, and more recently, a major entertainment center for the area. Only about 40 acres of wetlands remain in its northeast section. Project ReWild was proposed by the San Diego Audubon Society to rehabilitate that area. As part of the project, the University of San Diego would collect baseline geologic, biologic, and physio-chemical data on the bottom of the Bay (benthic zone) from 2015 to 2019. This information on the sediment, water temperature, and nutrients will help in the design of efforts to rehabilitate that wetland area.

HOPEFUL SIGNS

Birds

There are some very hopeful signs for the health of San Diego Bay. The California condor (*Gymnogyps californianus*) is a vulture and the largest land bird in North America. It is also an amazing story of recovery from near extinction. Their numbers plummeted in the 20th century. Poaching, lead poisoning from bullets that killed the carrion they feed on, and loss of habitat took a toll. In fact, it was completely extinct in the wild in 1987. The 27 living condors were captured in an effort to try to save the species at the San Diego Wild Animal Park and the Los Angeles Zoo. In 1991, they were reintroduced to the wild and have increased to 488 birds by 2018. The double-crested cormorant (*Phalacrocorax auritus*) is a common bird throughout North America, but in the late 19th century, their numbers plummeted. By the 21st Century, the population had increased (SDCBA, nd).

There have also been some success stories. For example, peregrine falcons (*Falco peregrinus*) were nearly wiped out by pesticides, but they have made a comeback along the San Diego coast. In other cities, they have thrived by preying on other birds mostly, such as rock doves (*Columba livia*). They nest in nesting in churches, window ledges of skyscrapers, and the towers of suspension bridges. However, this has yet to be reported in San Diego.

The California condor is another success story. The birds were nearly extinct, but a captive breeding program at the San Diego Zoo has brought them back. The last condor seen flying in San Diego was in 1910. Now hundreds of condors from the breeding program have been released in California, Arizona, and Mexico. A number of birds were releases into Baja California's Sierra San Pedro de Martir National Park, and recently some of those birds were near the U.S.-Mexico border. They have since expanded their territory into San Diego County.

Species Diversity and Stability

To know how well the various efforts are doing in removing pollution from San Diego Bay, it is important to know the number of species in the Bay. Sorensen et al. (2013) attempted to conduct a census of the living organisms in the Bay. They then compared their results to those from previous studies to determine how each species is doing. They also examined the sediments at the bottom of the Bay. Most of the invertebrates live in the sediments. In this rapid assessment study, the team looked at species richness, abundance, diversity, and assemblage. They collected and identified 6,477 organisms from 299 species and 13 phyla. The species found were similar to those reported in previous studies, but some differences were found in distributions. Finally, there did not seem to be a significant increase in invasive species.

Restoration Returns

Through the Partners for Fish and Wildlife Program and Coastal Program, the U.S. Fish and Wildlife Service seeks to complete conservation projects with the help of citizen and community groups (Laughland et al., 2013). The two programs provide funding for specific projects. An example of a successful project is the restoration of the salt ponds on San Diego Bay. That project sought to reverse the degradation of the shoreline of the Bay. By the end of the project, they had rehabilitated 300 acres of estuary habitat at three locations. Shorebirds moved in immediately. The project also was beneficial to the community. It provided $13.4 million to the local economy and 130 jobs.

RESTORING THE BAY

We cannot wave a magic wand and return the Bay to what it was 10,000 years ago or even 200 years ago. Now more than 3.5 million people live and work in the San Diego Bay region, and more are expected. The problem will get worse as more people need housing, jobs, transportation, and recreation. They will need freshwater and air and create waste. Wetlands are critical habitat. Each year more than 500,000 shorebirds and 700,000 waterfowl use the mudflats and salt ponds to rest and feed as they migrate along the Pacific Flyway. And beyond the needs of birds and other species are the needs of the growing human population in the Bay region. Some balance must be struck between the competing needs of those birds and the people who live in San Diego.

The good news is that there is a far greater realization of the importance of the environment and the value of wetlands and wild species than ever before. In addition, legislation now aims to protect the environment and to prevent pollution. It will take all of this and more to maintain the environment in the face of additional growth, sea level rise, and global warming.

REFERENCES

Auffhammer M, Kellogg R (2009) Clearing the air? The effects of gasoline content regulation on air quality. *American Economic Review* 101: 2687–2722.

Ayad M, Li J, Holt B, Lee C (2020) Analysis and classification of stormwater and wastewater runoff from the Tijuana River using remote sensing imagery. *Frontiers in Environmental Science* 8: 240.doi.org/10.3389/fenvs.2020.599030

Brandt JT, Sneed M, Danskin WR (2020) Detection and measurement of land subsidence and uplift using interferometric synthetic aperture radar, San Diego, California, USA, 2016–2018. *Proceeding of the International Society of Hydrological Sciences* 382: 45–49.

Carrillo CD (2004) Predator management for the protection of the endangered California least tern (*Sterna antillarum brownii*) and documentation of bullsnake (*Pituophis catenifer*) predation in San Diego County, California. in *Proceedings of the 21st Vertebrate Pest Conference*. RA Timm and WP Gorenzel, conference editors. University of California, Davis, USA, pp. 13–16.

Callaway JC, Zedler JB (2004) Restoration of urban salt marshes: Lessons from southern California. *Urban Ecosystems* 7: 107–124.

Capello M, Cutroneo L, Castellano M, Orsi M, Pieracci A, Bertolotto RM, Povero P, Tucci S (2010) Physical and sedimentological characterisation of dredged sediments. *Chemistry and Ecology* 26(sup1): 359–369.

Carse A, Lewis JA (2020) New horizons for dredging research: The ecology and politics of harbor deepening in the southeastern United States. *WIREs Water* e1485. https://doi.org/10.1002/wat2.1485.

Carson RT, Damon M, Johnson LT, Gonzalez JA (2009) Conceptual issues in designing a policy to phase out metalbased antifouling paints on recreational boats in San Diego Bay. *Journal of Environmental Management* 90: 2460–2468.

CDFW (2021) Marine invasive species. California Department of Fish and Wildlife. Retrieved from: https://wildlife.ca.gov/OSPR/Science/Marine-Invasive-Species-Program/Definition; accessed February 6, 2021.

Ciriminna R, Bright FV Pagliaro M (2015) Ecofriendly antifouling marine coatings. *ACS Sustainable Chemistry & Engineering* 3: 559–565.

de Jesus Piñon-Colin T, Rodriguez-Jimenez R, Rogel-Hernandez E, Alvarez-Andrade A, Toyohiko Wakida F (2020) Microplastics in stormwater runoff in a semiarid region, Tijuana, Mexico. *Science of the Total Environment* 704: 135411.

Dooley EC (2019) Boot-melting Mexican sewage has San Diego seeking help. *Bloomberg Law*. Retrieved from: https://news.bloomberglaw.com/environment-and-energy/boot-melting-mexican-sewage-has-san-diego-seeking-help; accessed February 7, 2021.

EPA (2019) Air Quality. Cities and Counties. *Environmental Protection Agency*. Retrieved from: https://www.epa.gov/air-trends/air-quality-cities-and-counties; accessed February 19, 2021.

Fraser M, Short J, Kendrick G, McLean D, Keesing J, Byrne M, Caley MJ, Clarke D, Davis A, Erftemeijer P, Field S, Gustin-Craig S, Huisman J, Keough M, Lavery P, Masini R, McMahon K, Mengersen K, Rasheed M, Statton J, Stoddart J, Wu P (2017) Effects of dredging on critical ecological processes for marine invertebrates, seagrasses and macroalgae, and the potential for management with environmental windows using Western Australia as a case study. *Ecological Indicators* 78: 229–242.

Gersberg RM, Daft D, Yorkey D (2004) Temporal pattern of toxicity in runoff from the Tijuana River Watershed. *Water Research* 38: 559–568.

Karner A, Eisinger D, Bai S, Niemeier D (2009) Mitigating diesel truck impacts in environmental justice communities: Transportation planning and air quality in Barrio Logan, San Diego, California. *Transportation Research Record* 2125: 1–8.

Larson RN, Morin DJ, Wierzbowska IA, Crooks KR (2015) Food habits of coyotes, gray foxes, and bobcats in a coastal southern California urban landscape. *Western North American Naturalist* 75(3): 10.

Laughland D, Phu L, Milmoe J (2013) Restoration Returns. U.S. Fish and Wildlife Service. Retrieved from: https://www.fws.gov/home/pdfs/restoration-returns.pdf; accessed February 11, 2021.

Ludka BC, Guza RT, O'Reilly WC (2018) Nourishment evolution and impacts at four southern California beaches: A sand volume analysis. *Coastal Engineering* 136: 96–105.

Malone KQ (2020) San Diego and Tijuana's shared sewage problem has a long history. *Washington Post.* Retrieved from: https://www.washingtonpost.com/outlook/2020/06/02/san-diego-tijuanas-shared-sewage-problem-has-long-history/; accessed February 7, 2021.

Meckstroth AM, Miles AK (2005) Predator removal and nesting waterbird success at San Francisco Bay, California. *Waterbirds* 28: 250–255.

Milligan B, Holmes R (2017) Sediment is critical infrastructure for the future of California's Bay-Delta. *Shore & Beach* 85(2).

Moreno-Mateos D, Alberd A, Morriën E, van der Putten WH, Rodríguez-Uña A, Montoya D (2020) The long-term restoration of ecosystem complexity. *Nature Ecology and Evolution* 4: 676–685.

Morin DJ, Higdon SD, Holub JL, Montague DM, Fies ML, Waits LP, Kelly MJ (2016) Bias in carnivore diet analysis resulting from misclassification of predator scats based on field identification. *Wildlife Society Bulletin* 40: 669–677.

Morzaria-Luna HN, Zedler JB (2007) Does seed availability limit plant establishment during salt marsh restoration? *Estuaries and Coasts* 30: 12–25.

Mossman HL, Davy AJ, Grant A (2012) Does managed coastal realignment create salt-marshes with "equivalent biological characteristics" to natural reference sites? *Journal of Applied Ecology* 49: 1446–1456.

Reed SE, Larson CL, Crooks KR, Merenlender AM (2014) Wildlife response to human recreation on NCCP reserves in San Diego County. Wildlife Conservation Society Agreement No/LAG #: P1182112.

Reilly ML, Tobler MW, Sonderegger DL, Beier P (2017) Spatial and temporal response of wildlife to recreational activities in the San Francisco Bay ecoregion. *Biological Conservation* 207: 117–126.

Ruiz-Jaen MC, Aide TM (2005) Restoration success: How is it being measured? *Restoration Ecology* 13: 569–577.

SDBNWR (nd) Final Predator Management Plan - San Diego Bay National Wildlife Refuge. Retrieved from: https://www.fws.gov/uploadedFiles/Region_8/NWRS/Zone_1/San_Diego_Complex/San_Diego_Bay/Sections/What_We_Do/Conservation/PDFs/Volume_2/Appendix%20M.pdf; accessed February 12, 2021.

SDCBA (nd) San Diego County Bird Atlas. University of San Diego. Retrieved from: https://map.sdsu.edu/group2007Spring/group2/Cormorants.htm#:~:text=The%20Double%2Dcrested%20is%20our,have%20formed%20only%20since%201988; accessed February 6, 2021.

Sorensen K, Swope B, Kirtay V (2013) *Marine Ecologic Index Survey of San Diego Bay*, Tech. rep., SSC Pacific, San Diego, CA. Retrieved from: https://apps.dtic.mil/sti/pdfs/ADA603815.pdf; accessed February 11, 2021.

Steiner AL, Tonse S, Cohen RC, Goldstein AH, Harley RA (2013) Influence of future climate and emissions on regional air quality in California. *Journal of Geophysical Research: Atmospheres* 111: D18303.

Stralberg D, Brennan M, Callaway JC, Wood JK, Schile LM, Jongsomjit D, Kelly M, Parker TV, Crooks S (2011) Evaluating tidal marsh sustainability in the face of sea-level rise: A hybrid modeling approach applied to San Francisco Bay. *PLoS One* 6(11): e27388.

Stumpner EB, Kraus TEC, Liang YL, Bachand SM, Horwath WR, Bachand PAM (2018) Sediment accretion and carbon storage in constructed wetlands receiving water treated with metal-based coagulants. *Ecological Engineering* 111: 176–185.

Svejkovsky J, Nezlin NP, Mustain NM, Kum JB (2010) Tracking stormwater discharge plumes and water quality of the Tijuana River with multispectral aerial imagery. *Estuarine Coastal and Shelf Science* 87: 387–398.

Tinkler T, Ahearne M, Schumann MJ (2019) Collaborative species and habitat conservation efforts in San Diego County: A systematic needs assessment to guide the San Diego End Extinction Initiative. *Environment 1*. Retrieved from: https://digital.sandiego.edu/npi-environment/1/; accessed February 18, 2021.

USD (nd) Air Quality. How are we doing? University of San Diego. Retrieved from: https://www.sandiego.edu/soles/hub-nonprofit/initiatives/dashboard/air-quality.php, February 19, 2021.

Wilber D, Clarke D (2010) Dredging activities and the potential impacts of sediment resuspension and sedimentation on oyster reefs. *Proceedings of the Western Dredging Association Technical Conference*, June 6–9, 2010, San Juan, Puerto Rico, USA, pp. 61–69.

Yee D, Wong A (2019). *Evaluation of PCB Concentrations, Masses, and Movement from Dredged Areas in San Francisco Bay*. SFEI Contribution #938. San Francisco Estuary Institute, Richmond, CA.

Zedler JB, Lindig-Cisnerow R (2000) Functional equivalency of restored and natural salt marshes. In: *Concepts and Controversies in Tidal Marsh Ecology* (eds. Weinstein MP, Kreeger DA) Springer, New York, pp. 565–582.

10 Future of the Bay

The time that it has taken to build San Diego Bay is hard for most of us to imagine. Our lives are busy, and we see events in terms of years and decades. Even centuries are beyond our firm grasp.

The physical structure of San Diego Bay has remained relatively stable for the last few million years. Sea levels have risen and fallen over that time so that specific areas were either dry or underwater. The volcanoes are long extinct. With the exception of the coming and going of seawater, nature has more or less left the Bay as it is. For the last 200 years, the major force for change has been human activity.

Of course, this stasis cannot continue forever. Large-scale events are not only possible, but certain, and nature has tens of millions of years to play with. The same forces that created the San Diego Bay continue to be active, but just not on a time scale that humans can comprehend.

Periodically, an earthquake occurs to remind us of where we live. The last large-scale wake-up call was in 1862, when an earthquake measuring 6.0 hit the Rose Canyon Fault. That fault is estimated to have a major earthquake every 700 years, and that earthquake could be as large as a 6.7 magnitude. In 1862, San Diego was sparsely populated. Today millions of people live in the city, and a major earthquake would be devastating. More than 100,000 homes could be damaged, roads and bridges would fail, gas and water service could be out for months, and the Mission Bay area might settle 30.5 cm.

Eventually, the tectonic forces that still push on the North American and Pacific Plates will cause earthquakes, and some of them will be devastating. Sea levels are again rising, and atmospheric rivers, droughts, landslides, and fires will peck away at the Bay Area. In aggregate and over enough time, the Bay will be unrecognizable.

PEOPLE AND MORE PEOPLE

San Diego Bay is a beautiful region with a mild Mediterranean climate, little rain, and a wonderful shoreline, mountains, and deserts. It has a world-class harbor and serves as a gateway between Mexico and the United States. For all of these reasons, it is a highly desirable place to live and work, and thus, the population is likely to continue to grow over the next decades. The natural setting around the Bay is also fragile, and the pressures of more humans will stress all of those natural elements that are so desirable.

The arrival of humans started a series of changes. Those changes began very slowly. Native Americans have been in the San Diego Bay Area for over

DOI: 10.1201/9780429487460-10

10,000 years. Since then, Spanish colonists and later Europeans and finally Americans and others arrived. In particular, the arrival of Europeans and Americans brought huge modifications to the shape of the Bay, the wetlands that surround the Bay, and the quality of the water and air in the area. A significant portion of the wetlands and tidal flats were filled and built on. More than 75% of the wetlands in the Bay were lost (Brophy et al., 2019). Rivers were channeled. And a wide variety of pollutants were released into the Bay water. Humans also degraded the air and introduced new alien species into the Bay that threaten the native species. Among the many are Asian kelp (*Undaria pinnatifida*), scarlet pimpernel (*Lysimachia arvensis*), hollyhock rust (*Puccinia malvacearum*), Japanese wireweed (*Sargassum muticum*), and goldspotted oak borer (*Agrilus auroguttatus*). Many of these arrived with imports in or on shipping. Some were released intentionally.

The population of the San Diego Metro Area is now over 3.3 million, and by 2050, it will be over 4.0 million (SANDAG, 2010). Fortunately, in the last few decades, there has been increasing awareness of the need to maintain and reclaim some of the lost wetlands as a defense against sea level rise and to protect wildlife in the surrounding areas. Nevertheless, the increasing population will require housing, water, sanitation, transportation, recreation, and much more. The infrastructure to support all those new people will stress the natural Bay.

FUTURE LOSS OF SPECIES

Tragically, many species are being lost, and it is far from clear that this trend can be slowed. Many examples can be cited. Nearly 75% of flying insects have been lost over the last 27 years (Hallmann et al., 2017). The loss of honey bees has been prominent in the news (Hristov et al., 2020). A large number of birds are also being lost (Loss et al., 2015). In another analysis, Mason et al. (2021) compared the potential range of a species with its actual range today. They used computer modeling to determine the range of birds considered not to be threatened in Great Britain. They found that 42% of the birds have wider ranges than expected and 28% have narrower ranges. These observations could help with attempts to maintain healthy bird populations. Finally, amphibians and particularly frogs have suffered enormous losses in recent years for unknown causes (Allentoft and O'Brien, 2010). Many scientists have begun to wonder if the Earth is in the midst of a sixth mass extinction. The Earth has so far survived five mass extinctions in which very large numbers of species were lost in geologically short periods of time. The most famous mass extinction killed off the dinosaurs. The rate of loss of species now certainly seems to suggest that another event is underway, and this time, the cause is clear. It is human activities (e.g., development, habitat loss, and partitioning) and their major product: climate change.

The global loss of species is also being felt in Southern California and in the San Diego Bay region. The rapid growth of population in Southern California continues to cause land to developed for housing and jobs. Riordan and Rundel (2014) examined the effects of land usage and climate change on California sage scrub (*Salvia* spp) by modeling their distribution under unlimited and spatially limited projections. They looked at 20 main species of sage scrub that are common in coastal central and southern California and into Baja California. The models depended greatly on the dispersal criteria, and land use was more important than climate change in Central areas, but both were important in Southern California. They conclude that both factors must be considered in future efforts to maintain species diversity.

Even seemingly innocuous activities can be disruptive of habitats. Dugan and Hubbard (2010) showed that beach grooming can contribute to a change of coastal strand ecosystems to plain sand beaches. The loss of plants along the coasts will result in the loss of sediment, dunes, and diversity even as sea levels are rising, and the need for those buffer zones is even greater.

EROSION

Much of the area of cities and the surrounding suburbs are covered with concrete and asphalt. The natural flow of water from rain has been disrupted and redirected into storm sewers. Creeks and rivers that previously flowed freely to the Bay or ocean become channeled. Water movement and erosion in these areas have been well studied and can even be divided into stages. As urban areas develop, large amounts of sediment are initially mobilized as building occurs (Chin, 2006). The amount can be 2–10 times more than before. That slows down as building is completed. While those measures contain the water and the erosion, some researchers believe that a degree of river bank erosion is desirable. Florsheim et al. (2008) point out that such erosion encourages a healthy vegetation succession and provides dynamic habitat for plants and animals.

Throughout the California coast, many communities have been built on uplifted marine terraces, and those terraces have long been subject to erosion by ocean waves. In fact, 86% of those coastal cliffs are actively eroding (Griggs, 1995). The cliffs in San Diego County are steep and 5–115 meters high. Benumof et al. (2000) examined erosion on sea cliffs caused by wave action. They measured the force of the waves at several heights above the water. They found that the wave energy was actually secondary to the material composition of the cliff in determining the rate of erosion.

Olsen et al. (2008) used a terrestrial light detection and ranging (LIDAR) system and interactive visualization techniques to study sea cliff erosion. By frequently monitoring cliffs, they found that they could gain significant insights into how cliff erosion functions. Young et al. (2010) used this same technique to examine 400 meters of sea cliffs near Del Mar during the rainy season by airborne and terrestrial lidar. Both methods worked well, but the terrestrial lidar was found to be much more accurate. However, the airborne method was much more rapid and thus could cover more territory. Olsen et al. (2016) found that the material that is eroded from the cliff face has beneficial functions. First, it protects the remaining cliff face from further damage as it takes the brunt of wave action. Second, it adds to the beach material as it is broken down over time.

LANDSLIDES AND SUBSIDENCE

Landslides

Climate change will bring more extreme weather events, and landslides are sure to follow those heavy rains. Mud and silt flow downhill, and humans activities associated with development and agriculture disrupt the land and make it more susceptible to movement. For example, clearing land for crops encourages erosion that fills rivers and streams. In more recent times, deforestation and urban development have resulted in more erosion, while dams have stopped the natural movement of the material (Voosen, 2020). Landslides cost lives and damage property. Since landslides are often associated with extreme weather events, Cordeira et al. (2019) compared landslides from 1871 to 2012 with records of Pacific winter storms and atmospheric river events. They found that 76% of the landslides occurred during storms and 82% occurred during an atmospheric river.

Of course, not all landslides happen quickly. Slow-moving landslides are a problem for much of California. They rarely cost lives, but they can be extremely destructive to structures and infrastructure. The slides depend on the soil, climate, and earthquake activity of the area, but the mechanisms that initiate and maintain them are not well understood. They often occur in soils rich in clay and rock that are mechanically weak and have high levels of seasonal precipitation. The California Coast Ranges are an ideal place to study slow landslides. Lacroix et al. (2020) reviewed the forces that control slow landslides in this environment. Those factors include the overall geology, climate, and tectonics and also precipitation and groundwater, earthquakes, river erosion, anthropogenic activities, and external material supply. Rivers can block slow-moving landslides, and landslides can block rivers. Finnegan et al. (2019) examined the several effects on the blockage of rivers by landslides and found that wider rivers are less affected by landslides. More narrow streams can be completely blocked by the landslide. Also, rivers vary in their ability to mobilize the material in the landslide. Land that has been moving slowly can suddenly fail with serious implications for people and property. Increasing pore pressure by fluids in the soil increase creep rates and can reach a stochastic point at which the system fails (Agliardi et al., 2020).

Subsidence

San Diego obtains 90% of its freshwater from distant sources. Only 10% comes from local rivers. The city has invested in desalination plants to purify brackish water from a coastal aquifer. However, pumping out groundwater has potential implications for subsidence. Brandt et al. (2020) used interferometric synthetic aperture radar to study the effects of subsidence before and after periods of drawing underground water. They found subsidence of about 75 mm after pumping. However, much of that was regained (45 mm) after the groundwater had been replenished.

CLIMATE CRISIS

The climate crisis will affect many ecosystems in the San Diego region. Habitat will be lost and fragmented. The uses of land will change, and fires will be more common and intense. Estimates are that the region will warm by 4–9 F and precipitation will decrease by 15–25%. However, periods of rain or drought will become more extreme. Because the area has many microclimates, the effects will be felt more by plants and animals. The Santa Ana winds will continue to increase the threat of fire.

TOO MUCH WATER

On October 2, 1858, San Diego was hit by a tropical cyclone that lasted more than 24 hours (Chenoweth and Landsea, 2004). Hurricane-level winds caused severe damage, and heavy rains fell along the coast and inland. Damage was wide spread, but the population was only around 4000 at the time. The remnants of tropical cyclones often reach Southern California, particularly in years of El Niños. However, this is the first example of a hurricane-force storm actually reaching the area, and records show no evidence of an El Niño that year.

In the 20th Century, four tropical storms hit the San Diego Area. With the accelerating global warming now occurring, more anomalous weather events are likely. Relatively rare events, such as the hurricane, might become more common in the future. In any case, it is clear that powerful forces can and do affect the region, and they will continue to do that in the future.

Atmospheric Rivers

California's mild Mediterranean climate makes the state unusually sensitive to drought and flooding. The amount of rain varies significantly from year to year. For example, the year 1862 was extremely wet, but the years after that featured severe droughts. Some studies have shown that the state will experience more very wet years in the future (Swain et al., 2018).

Atmospheric rivers bring very large amounts of rain to California and have been responsible for breaking some of the state's worst droughts (Figure 10.1). Thus, understanding these events is critical to managing water reserves in the state. Gershunov et al. (2019) examined 16 global climate models to determine how they predict atmospheric rivers. The five most accurate models confirm the variability in precipitation in California and most of the Western United States. Interestingly, they found that any increase in rainfall is due almost entirely to atmospheric rivers. However, O'Gorman (2015) points out that the effects of precipitation extremes are not simple. Some factors, such as convection effects and the duration and type of precipitation, are not well understood. Climate change contributes another confounding factor to the mix. California only receives rain on about 5–15 days per year (Dettinger et al., 2011). Some of the storms can be quite big and involve atmospheric rivers, contributing 20–50% of

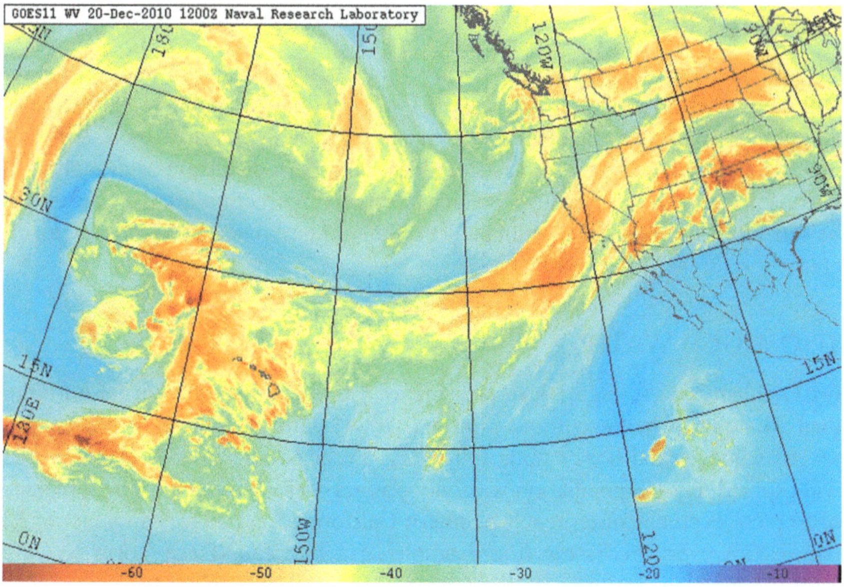

FIGURE 10.1 **Atmospheric rivers**. In some cases, long narrow bands of water vapor form. They can stretch for thousands of miles, and when they reach California, they can release enormous amounts of rain in a very short time. The image is from the United States Naval Research Laboratory, Monterey.

the annual precipitation. Those storms can be the difference between a year of plenty or one of drought.

As the atmosphere warms due to climate change, it can hold more moisture and thus increase the amount of rain carried in atmospheric rivers. In recent years, the increase has been 7% per degree of surface temperature rise (Algarra et al., 2020). Based on these data, Algarra et al. predict an increase in the amount of rain that will make landfall during atmospheric river events. Shields and Kiehl (2016) used the Community Climate System Model to predict future atmospheric activity on the West Coast. They found that these events will increase in intensity with global warming, particularly in Southern California.

Global Warming

Climate change is nothing new. It has occurred throughout the history of the Earth. Periods of hotter and colder temperatures have alternated, and the sea levels have risen and fallen. The difference now is that human activities are accelerating global warming. The use of fossil fuels releases large amounts of carbon dioxide into the atmosphere. That and other greenhouse gases trap heat, the surface albedo decreases, and even more heat energy is trapped in the atmosphere. This positive feedback mechanism continues to accelerate the cycle.

FIGURE 10.2 Sea level rise. As climate change warms the planet, the polar ice caps and permanent glaciers are melting. As more ice melts, sea levels rise, and low-lying areas in the San Diego region are at risk of flooding. This projection is based on a 1.5-meter sea level rise in combination with possible 100-year return period coastal storms. (Illustration courtesy of US Geological Survey).

As the polar caps and glacier ice melt, the level of the oceans rises, and coastal communities will be more and more threatened with inundation by flooding by storm and tidal surge. The San Diego Area includes a many low-lying areas that contain valuable housing and infrastructure, including the airport and port facilities (Figure 10.2). Global warming will greatly enhance the threat to low-lying areas around San Diego Bay and throughout the region. As sandy beaches are lost, significant housing and infrastructure will be even more exposed (Zhang et al., 2004). This will present enormous challenges to authorities for planning and prevention of damage. The problem is exacerbated by the fact that much of the land around the Bay is on fill. One estimate is that, by 2100, nearly 6000 homes will be at risk with a combined value of nearly $6 billion (UCS, 2018). The potential disruption to communities and industries is considerable and will require significant planning by governments. Barriers (e.g., dikes, levees, and sea walls) can be constructed to protect some structures. Others may have to be moved. The cost will be staggering. Governmental agencies will have to rethink development in low-lying areas, and insurance companies will need to reevaluate risk for development.

Climate change is upon us today, and it will only get worse without serious action to reverse the warming. Kopp et al. (2014) predicts that sea levels will rise by as much as 1.2 meters between 2000 and 2100. Other scientists estimate that, by 2200, our atmosphere will contain higher levels of carbon dioxide than any time over the last 650,000 years. The warming will accelerate the melting of the polar ice caps and cause the levels of the oceans to rise even faster.

The Rahmstorf 2007 semiempirical method is often used to estimate sea level rise (Rahmstorf, 2007). It links sea level rise to global mean temperatures. Messner et al. (2011) used this model, along with wave data, tides, weather effects, El Nino effects, and longer-term sea level changes to map areas of San Diego that might be threatened by sea level rise. They found six low-lying areas that are particularly at risk (e.g., South Imperial Beach, Coronado Beach and Shores, Mission Beach, South La Jolla Shores, North Del Mar, and Oceanside Harbor). In California, coastal base flood elevations range from 10.5 meters in Mendocino County in the north to 2.3 meters in San Diego Harbor in the south (Heberger et al., 2011). Global warming will make this much worse.

Not only developed areas will be affected. Marsh and tidal flats will be among the earliest victims. Up to 95% of the marsh area will be inundated, and its plant communities will be lost (Takekawa et al., 2013). Those tidal wetlands are critical to the overall health of the Bay. They filter water, slow floods, protect infrastructure, and sequester carbon. Some might be saved by adding material to the marshlands, but ultimately, they will likely be flooded. In the natural world, those plants might migrate with the rising water, but most current marsh areas are bounded by developed areas and so migration is not possible. In recent decades, more and more efforts have been made to preserve and even expand wetlands around the Bay (see Chapter 9). However, those efforts will have to be greatly accelerated if they are to keep pace with sea level rise. And the cost will be huge.

Global warming is a threat to many organisms, and its effects are already being seen in plants and animals in San Diego Bay and surrounding areas. Every group is affected, but some more than others. Nearly a third of all amphibians are at risk today (Wake and Vredenburg, 2008). Loss of habitat, pollution, exotic species, and disease amplifies the effects of climate change. Birds are also quite threatened. Since 1970, we have lost 3 billion birds or 29% of all birds (Rosenberg et al., 2019). Fish are stressed by dams and other infrastructure projects, pollution, and invasive species. Nearly 40% of all North American freshwater and anadromous fish are at risk (Jelks et al., 2008). Invertebrates account for 97% of all species. Bees are critical for the pollination of a large majority of human food crops, but they have been hard hit in recent years (Klein et al., 2018). Many plants are also under pressure, and they cannot easily migrate to more suitable environments (Tilman et al., 1994).

TOO LITTLE WATER

Unfortunately, the opposite also occurs. San Diego suffers from too little water, and that back and forth between flood and drought is likely to continue to happen in the future. A look at climate changes that occurred in the last 1000 years might be helpful in thinking about what might happen in the future. Brunelle and

Anderson (2003) conducted such as study by examining fossil pollen and charcoal in sediments from Siesta Lake in California. They then used those data to build a record of climate and fire during the Holocene. Two periods were particularly informative. The "Mediaeval Warm Period" lasted from 950 to 1250 CE and featured temperatures that were similar to the early- to mid-20th Century. The early Holocene (beginning approximately 11,000 years ago) was a period of particularly strong sunlight that increased temperatures. They conclude that global warming will result in an increase in droughts and in the associated wildfires.

California has a long history of droughts. They tend to be triggered by too few winter storm and unusually high temperatures that result in very low soil moisture and available water. In 2012–2016, California suffered a very severe drought. Those years featured record high temperatures and very little precipitation. Although the drought ended, it almost certainly presages similar events in the coming years. Ullrich et al. (2018) attempted to model what those effects might be if such drought conditions were to occur in the years 2042–2046. They found that a mid-century drought would be far worse due to climate change. Temperatures will be much greater and for longer. The snowpack will be very low, and the soil will be dry. Many trees and other plants will die. The fire danger will also be much greater.

California's climate lends itself to periods of wet and dry. The transition from the severe drought of 2012–2016 to the very wet years of 2016–2017 is a good example of this variability. Swain et al. (2018) used the Community Earth System Model Large Ensemble of climate model simulations to examine what California's climate might be like in this century. They find that there will be more years of too much water and also those with too little.

Climate change is occurring now, but its effects have been around for too short a time to see many changes in the physical landscape. East and Sankey (2020) reviewed the literature to try to determine what changes to the geomorphology have and will yet occur in the western United States. These might be in the form of changes to slope stability, watershed sediment yields, river morphology, and wind erosion. Not all of these have been affected yet, but they did find evidence for changes to slopes and wind erosion. In the future, all of the effects are likely to be seen as climate change progresses, and these changes will present challenges to many aspects of modern life.

Southern California and San Diego are sensitive to drought. They depend on the over-taxed Colorado River and Northern California for freshwater supplies. With global warming, periods of drought are likely to be more common and longer. San Diego will have to look for additional options for water in the future. The area has invested in water desalinization plants to tap seawater and the brackish water contained in the San Diego Basin Aquifer. They are expensive, but the availability of seawater is unlimited for the city's needs. In addition, new policies by the various levels of government will be needed to encourage or even require conservation. Maggioni (2015) analyzed water usage and ordinances in place from 2006 to 2010. She found that water usage was reduced effectively by mandates to do that, but subsidies to purchase devices to limit supply or increased rates were not. Clearly, more imaginative policy will be needed to attempt to deal with the increasing demand for freshwater.

WILDFIRES

The 21st Century has seen an increase in the number and severity of wildfires in San Diego County. In 2003, the Cedar Fire burned over 273,000 acres. The Santa Ana winds caused the fire to spread rapidly. It destroyed over 2800 buildings and killed 15 people. It took more than 2 months to control the fire. In 2007, the Witch-Guejito and Harris Fires burned 288,000 acres and killed seven people. As global warming brings higher temperatures and affects levels of rainfall, more fires are expected.

The wildfires in 2007 burned more than 360,000 acres. Zauscher et al. (2013) examined the effect of these fires on the local air quality. They found that 84% of the 120–400-nm particles were biomass burning aerosols. Those particles absorb solar radiation and serve as nuclei for cloud condensation. They also are associated with adverse health effects. The make-up of the particles also varies according to the temperature of the fire. Hotter fires produce particles with more inorganic material and more soot. Cooler smoldering fires produce more organic carbon-rich aerosols.

The Santa Ana winds in the 2003 fires were not unusually strong, but the other weather components featured a multiyear drought, extensive numbers of dead trees, and chaparral (Westerling et al., 2004). In addition, late winter rains, a cool spring, and an early summer yielded a bumper crop of grasses that dried in the summer. Finally, years of fire prevention had increased the burden of flammable material in the wildlands beyond what it would have historically been. Thus, a "perfect storm" of events led up to a greater than normal fire season. The Santa Ana winds are foehnlike winds. They result from a mass of cool, dry air flowing from the interior basins and funneling through the Sierra Nevada Mountains to the coast. As the air sinks, it is compressed and warmed. The winds are typically 40–60 km/h, but can reach 100 km/h. Their relative humidity is very low, and so, these hot, dry fast winds are ideal for supporting fires.

The vegetation in San Diego County can be divided into six general classes according to the dominant plants. These include herbaceous, sage scrub, chaparral, hardwoods, conifers, and desert. In the 20th century, the trends suggest that chaparral and sage scrub burn most often, and the number of fires involving hardwoods and conifers burn is decreasing (Wells et al., 2004). They concluded that this is consistent with the increase in the number of fires where urban development intersects with wildland.

Land use planning and fire abatement measures can help in many cases (Syphard et al., 2013). However, that is expensive and often is in conflict with the desire of many to enjoy the benefits of living close to nature. However, the climate crisis and global warming, along with higher temperatures and stronger winds, are likely to further increase the number and intensity of wildfires and the threat to humans and their property.

A NEW ICE AGE?

About 20,000 years ago, the Earth was much cooler than it is today. Masses of water were frozen at the poles, and the ice sheet reached across much of North America. It covered all of Canada and New England and much of New York were covered by ice, as well as far south as Missouri. With all of that water locked up, sea levels were considerably lower, and the continents reached tens of kilometers past the current coasts. The sea level was so low that a land bridge formed between Siberia and Alaska, and humans migrated across that bridge to populate North America. That was the last Ice Age, but there have been several others. They occur about every 200 million years and last for some tens of millions of years. In fact, the last Ice Age is still upon us. We are simply in one of the warmer periods of that time.

Several theories attempt to explain the cause of the ice ages. One of the most intriguing comes from the great Serbian scientist Milutin Milanković. The Earth's orbit periodically changes from a circle to an ellipse, and the Earth tilts a bit more. These are small changes, but they affect the amount of sunlight hitting the Earths enough to cause the temperature variations that result in ice ages. In fact, their regularity can be used to predict the timing of the next Ice Age (Hays et al., 1976). The next Ice Age is estimated to begin in about 1500 years. Plate tectonics may also be involved. Changes to the arrangement of the continents may allow or preclude circulation of the oceans or the atmosphere. Changes in the amount of carbon dioxide in the atmosphere are another strong candidate.

Humans are a new complicating factor. We are pumping ever-larger amounts of greenhouse gases into the atmosphere (Maslin, 2016, 2020). Global temperatures are already rising at alarming rates. Could those rises be enough to offset the cooling that might naturally occur and postpone the next Ice Age? The answer is not clear.

Another possibility is a "little ice age." The sun also varies in its brightness. During a grand solar minimum, the total solar irradiance is reduced by 0.25%. That seems like a very small difference, but one of these, called the Maunder Minimum, cooled the Earth in 1650–1700. The connection is only theoretical. However, temperatures on Earth during that period were much cooler than today. Some scientists predict that such an event will occur between 2020 and 2070. Current scientific thought seems to indicate that this event will slow but not fully correct the influence of human activities on global warming.

PLATE TECTONICS AND EARTHQUAKES

The greatest force involved in building the San Diego Bay was the collision of the North American and Pacific Plates. As those two plates grind past one another, they generate earthquakes that deform the ground in three dimensions. Whole continents are changed and are continuing to change because those tectonic forces are still acting on the entire surface of the Earth. Opinions differ on the future, but some scientists believe that the Americans and Asia will join together to form the next supercontinent. Three computer models give that answer, but one (Mitchell et al., 2012) suggests that the Caribbean Sea and the Arctic Ocean will disappear, and South America will wind up alongside the eastern coast of North America.

How might that happen? Earthquakes in San Diego are mostly small, but some can be very large. Large quakes seem rare to us on our human time scale. However, they are actually quite common on a geologic time scale, and their cumulative effect can result in very large changes in geography. For example, in about 100 million years, "San Francisco" will be just off the west coast of Canada and nearing Alaska.

The most dangerous fault in the San Diego area is probably the Newport-Inglewood/Rose Canyon fault system. It runs right through the most populated parts of the area. The system has four main fault strands that are separated by three main stepovers of 2 km or less (Sahakian et al., 2017). Ruptures can initiate on one strand and be transferred to another strand at the stepover. So the energy of an earthquake is related to where it is initiated and the stepover. However, their calculations still indicate that a break on the whole length of the fault system could produce an earthquake of magnitude 7.3 or even 7.4.

Most people who live on the West Coast have experienced earthquakes. They seem uncommon, but that is a result of how short human lives are in comparison to the Earth. In fact, earthquakes are really very common. Becker and Geschwind (2016) posted a dramatic demonstration of the major earthquakes around the world from 2001 to 2015. In their work, the "ring of fire" that circles the Pacific Ocean and the other major faults and rifts are clearly shown by the number of earthquakes on them. The video provides a far better appreciation of the movement along those faults than any words. It's also easy to see that the West Coast is part of the "ring of fire."

For California, the "star" of the show is the San Andreas fault system. This continental transform fault runs about 1300 km from Southern California near the Salton Sea and to nearly Eureka in Northern California. The fault is the boundary between the North American and the Pacific plates. The Pacific plate is moving northwest and continues to grind against the North American plate, which is moving southwest. The movement is slow to us. In some areas, the sliding is not so smooth. The two masses can catch on each other. When that happens, the pressure can build up until it is suddenly released, and we feel that as an earthquake. The San Diego Bay region lies to the west of the San Andreas fault. That means that San Diego sits on the Pacific Plate. Earthquakes on the southern portion of the fault might be felt in San Diego. More importantly (in a geologic sense), San Diego will be affected by the overall movements on the fault. Christopher R. Scotese, a geologist at the University of Texas, Arlington, has calculated the results of future movements on the California coast. His fascinating website (www.scotese.com) contains projections of the Earth over the next 250 million years.

There has already been a great deal of land movement on this fault. For millions of years, the California coast has been migrating northward as part of the Pacific plate. The Pinnacles National Park just east of Monterey, California, is a great example. The tall rock formations are a favorite of rock climbers. They are the remnants of a large volcano. The softer rock around has eroded away, leaving the columns of very hard volcanic rock. Amazing, this is only half of the original volcano. The other half is still just east of Los Angeles in Southern California on the North American plate.

As mentioned above, geologists have begun to make some interesting predictions about the future of California. Clearly, these are highly speculative, and the farther into the future they go, the more speculative they become. However, their predictions of the next 50 million years are intriguing. Robert S. Dietz of the US Geologic Survey did some of the early work that validated plate tectonics on the ocean floor. In 1970, he published his predictions about the future of California in an interesting article in *Scientific American*. Based on the movements along the San Andreas fault, he predicted that, in 10 million years, Los Angeles will be repositioned to the current San Francisco Bay Area and, in 50 million years, it will be part of Alaska. San Diego would, of course, be just south of the relocated Los Angeles. More recent work by geologists has validated the general lines of Dietz's projections.

The area around the Salton Sea figures into these calculations. It contains the Salton Trough, a graben or a basin created by two parallel faults. The graben becomes depressed as the land on the two outsides rises. It resulted from the stretching as the San Andreas fault and the East Pacific Rise moved. The San Andreas and the Gulf of California Rift Zone both end near the south end of the Salton Sea.

The San Andreas fault is important to California's geologic future, but models premised on it are not without limitations. McCrory et al. (2009) modeled the subduction margins in Southern California and the slab windows that opened in that area. They cataloged the volcanic events that accompanied those slab windows from 28.5, 19, 12.5, and 10 million years ago. With these, they developed a model to describe the evolution of the continental margin. They then ran their model into the future to see what would happen. At 2 million years in the future, the bend in the San Andreas fault in Southern California becomes a problem. The Peninsular Ranges and the Sierras have been converging since 12.5 million years ago, but by 2 million years in the future, they will collide directly, and that pressure must be relieved somehow. McCrory et al. (2009) suggest that this stress could be relieved by new faults breaking through the lower Sierra Madre or by a shift of the movement between the plates from the San Andreas to the Walker Lane system.

Other scientists agree and have suggested alternate fates for California. The San Andreas fault is certainly still in the mix, but other weak points in the North American plate might also be involved. Interestingly, those factors are active at the Southern end of the San Andreas fault, east of San Diego. A rift in the Gulf of California forms a highly geologically active area that is breaking off Baja California from Mexico. Eventually, that rift will open deeper into Mexico and then California to form an inland sea.

Others see more possibilities. For example, in the next 8–10 million years, the rift in the Gulf of California might expand into the Walker Lane and extend northward to Lake Tahoe. Walker Lane is on the California-Nevada border and runs north of Mount Lassen to Death Valley and beyond where it connects to the San Andreas fault system (Faulds et al., 2005). Those scientists favor this model over the San Andreas fault because the fault has a long curve to the west that is not favorable for movement. To them, the more natural path is up Walker Lane. They believe that the San Andreas fault will become much less active as the movement is taken up by this new system. If they are correct, the rift that formed the Gulf of California will continue north on a path that is similar to that in Baja California. The path includes the Salton Sea, Mono Lake, Lake Tahoe, and Pyramid Lake and also a number of volcanoes. Wesnousky (2005) published an extensive study of the many faults in the Walker Lane. The westward bend of the San Andreas fault in Southern California limits movement on that fault system. The faults in the Walker Lane are far more linear and might offer a more optimal path to accommodate movement of the two plates.

The evidence supporting the Walker Lane hypothesis is growing. First, the fault system in that area already accounts for 15–25% of the movement along the North American and Pacific plates. This is a surprising amount, considering that the San Andreas fault had been assumed to be the dominant interface between the plates. Second, Baja California separated from the North American plate about 7 million years ago to form the Gulf of California. A chain of volcanoes warmed the continental crust, causing it to soften and creating a series of weak spots that allowed the land to separate. Third, about 13,000 years ago, Pyramid lake was part of an inland sea called Lake Lahontan. The lake is near a number of newly discovered faults, including the Pyramid Lake Fault, the Honey Lake Fault, and the Warm Springs Valley Fault. These developing faults might form part of the northern part of the new rift.

The rift in the Walker Lane provides an excellent opportunity to study a rift on land (Putirka and Busby, 2011; Busby, 2013). The fault system extends far beyond Lake Tahoe in the north. Interestingly, the Gorda, North American, and Pacific plates form a triple junction at Cape Mendocino in Northern California. The Gorda plate is a subplate of the Farallon plate that broke off as the Farallon plate was subducted under the North American plate. The rift in Walker Lane has moved northward as the Mendocino triple junction has also moved northward. Ultimately, the rift might break towards the triple junction.

Meldahl (2015) suggested three potential futures for Southern California. In all three, the Pacific plate continues to slip northward along the North American plate (Figure 10.3). The difference in the three futures is in where the major boundary is between the plates. In the first, more traditional view, the Pacific plate and that part of California on that plate continue moving northward. In the second, the movement is transferred from the San Andreas fault to Walker Lane as that rift continues to extend northward. In this scenario, a much larger portion of California moves northward, including the entire San Diego Bay Area. In either case, San Diego and California will be unrecognizable.

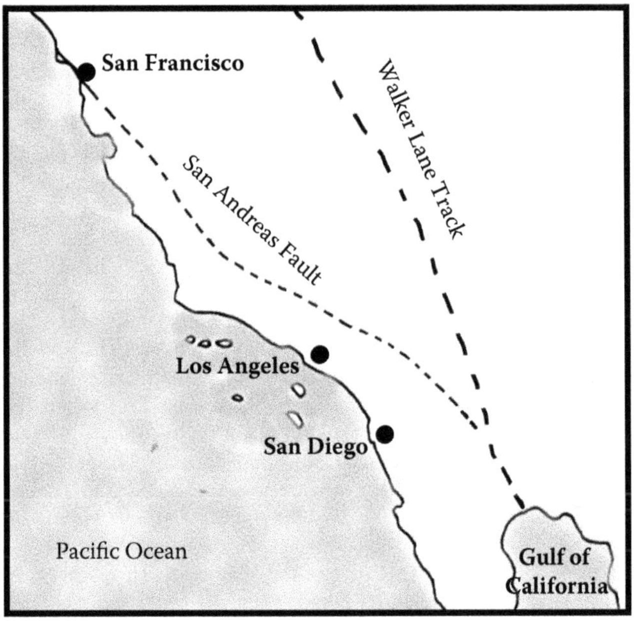

FIGURE 10.3 Three futures for California. The forces that built the current California are still at work and eventually will tear it apart. The San Andreas fault runs from east of San Diego to just south of San Francisco, where it goes to sea. In the first model of the future, the part of California west of the San Andreas fault continues to move northward eventually bringing San Diego nearly to the latitude of San Francisco. In the second, the stress moves to the Gulf of California, and the rift opens through the Walker Lane to north of Lake Tahoe. In the third, the split occurs farther east through central Nevada, where the crust is already stretched thin.

CONCLUSIONS

One might conclude from this chapter that the future of the San Diego Bay region is bleak. Certainly, challenges await the region. Two are probably the most serious. Tectonic movements are impossible to prevent. Those changes will happen, and the region will be changed beyond recognition. However, those changes are incredibly slow and, thus, not anything to worry about. Of course, a devastating earthquake could occur at any time on the San Andreas or Rose Canyon faults. The results could be profound with many deaths and billions of dollars in damages. The real threat is climate change. These days, many things are labeled as an existential threat. Climate change truly is an existential threat. And it will manifest as many crises: extreme rain and drought, winds, and wildfires. These will put a great deal of pressure on existing systems for water, food, and land as sea level rise. All of these will require significant amounts of planning and policy from all levels of government.

REFERENCES

Agliardi F, Scuderi MM, Fusi N, Collettini C (2020) Slow-to-fast transition of giant creeping rockslides modulated by undrained loading in basal shear zones. *Nature Communications* 11: 1352.

Algarra I, Nieto R, Ramos AM, Eiras-Barca J, Trigo RM, Gimeno L (2020) Significant increase of global anomalous moisture uptake feeding landfalling Atmospheric Rivers. *Nature Communications* 11: 1–7.

Allentoft ME, O'Brien J (2010) Global amphibian declines, loss of genetic diversity and fitness: A review. *Diversity* 2: 47–71.

Becker N, Geschwind LR (2016) Earthquakes—2001–2015. Science on a Sphere. *National Oceanic and Atmospheric Administration.* http://sos.noaa.gov/Datasets/dataset.php?id=643

Benumof BT, Storlazzi CD, Seymour RJ, Griggs GB (2000) The relationship between incident wave energy and seacliff erosion rates: San Diego County, California. *Journal of Coastal Research* 16: 1162–1178.

Brandt JT, Sneed M, Danskin WR (2020) Detection and measurement of land subsidence and uplift using interferometric synthetic aperture radar, San Diego, California, USA, 2016–2018. *Proceedings of the International Association of Hydrological Sciences* 382: 45–49.

Brophy LS, Greene CM, Hare VC, Holycross B, Lanier A, Heady WN, O'Connor K, Imaki H, Haddad T, Dana R (2019) Insights into estuary habitat loss in the western United States using a new method for mapping maximum extent of tidal wetlands. *PLoS ONE* 14(8): e0218558.

Brunelle A, Anderson RS (2003) Sedimentary charcoal as an indicator of late-Holocene drought in the Sierra Nevada, California, and its relevance to the future. *The Holocene* 13: 21–28.

Busby CJ (2013) Birth of a plate boundary at ca. 12 Ma in the Ancestral Cascades arc, Walker Lane belt of California and Nevada. *Geosphere* 9: 1147–1160.

Chenoweth M, Landsea C (2004) The San Diego hurricane of 2 October 1858. *Bulletin of the American Meteorological Society* 85: 1689–1697.

Chin A (2006) Urban transformation of river landscapes in a global context. *Geomorphology* 79: 460–487.

Cordeira JM, Stock J, Dettinger MD, Young AM, Kalansky JF, Ralph FM (2019) A 142-Year climatology of Northern California landslides and atmospheric rivers. *Bulletin of the American Meteorological Society* 100: 1499–1509.

Dettinger MD, Ralph FM, Das T, Neiman PJ, Cayan DR (2011) Atmospheric rivers, floods and the water resources of California. *Water* 3: 445–478.

Dugan E, Hubbard DM (2010) Loss of coastal strand habitat in Southern California: The role of beach grooming. *Estuaries and Coasts* 33: 67–77.

East AE, Sankey JB (2020) Geomorphic and sedimentary effects of modern climate change: Current and anticipated future conditions in the western United States. *Reviews of Geophysics* 58: e2019RG000692.

Faulds JE, Henry CD, Hinz NH (2005) Kinematics of the northern Walker Lane: An incipient transform fault along the Pacific–North American plate boundary. *Geology* 33: 505–508.

Finnegan NJ, Broudy KN, Nereson AL, Roering JJ, Handwerger AL, Gennett G (2019) River channel width controls blocking by slow-moving landslides in California's Franciscan mélange. *Earth Surface Dynamics* 7: 879–894.

Florsheim JL, Mount JF, Chin A (2008) Bank erosion as a desirable attribute of rivers. *Bioscience* 58: 519–529.

Gershunov A, Shulgina T, Clemesha RES, Guirguis K, Pierce DW, Dettinger MD, Lavers DA, Cayan DR, Polade SD, Kalansky J, Ralph FM (2019) Precipitation regime change in Western North America: The role of atmospheric rivers. *Scientific Reports* 9: 9944.

Griggs GB (1995) California's coastal hazards. *Journal of Coastal Research*, Special Issue No. 12: Coastal Hazards, 1–15.

Hallmann CA, Sorg M, Jongejans E, Siepel H, Hofland N, Schwan H, et al. (2017). More than 75 percent decline over 27 years in total flying insect biomass in protected areas. *PLoS One* 12: e0185809.

Hays JD, Imbrie J, Shackleton NJ (1976) Variations in the Earth's orbit: Pacemaker of the Ice Ages. *Science* 194: 1121–1132.

Heberger M, Cooley H, Herrera P, Gleick PH, Moore E (2011) Potential impacts of increased coastal flooding in California due to sea-level rise. *Climatic Change* 109 (Suppl 1): S229–S249.

Hristov P, Shumkova R, Palova N, Neov B (2020) Factors associated with honey bee colony losses: A mini-review. *Veterinary Sciences* 7: 166.

Jelks HJ, Walsh SJ, Burkhead NM, Contreras-Balderas S, Díaz-Pardo E, Hendrickson DA, Lyons J, Mandrak NE, McCormick F, Nelson JS, Platania SP, Porter BA, Renaud CB, Schmitter-Soto JJ, Taylor EB, Warren ML, Jr. (2008) Conservation status of imperiled North American freshwater and diadromous fishes. *Fisheries* 33(8): 372–407.

Klein A-M, Boreux V, Fornoff F, Murepele A-C, Pufal G (2018) Relevance of wild and managed bees for human well-being. *Current Opinion in Insect Science* 26: 82–88.

Kopp RE, Horton RM, Little CM, Mitrovica JX, Oppenheimer M, Rasmussen DJ, Strauss BH, Tebaldi C (2014) Probabilistic 21st and 22nd century sea-level projections at a global network of tide-gauge sites. *Earth's Future* 2: 383–406.

Lacroix P, Handwerger AL, Bièvre G (2020) Life and death of slow-moving landslides. *Nature Reviews Earth and Environment* 1: 404–419.

Loss SR, Will T, Marra PP (2015) Direct mortality of birds from anthropogenic causes. *Annual Review of Ecology, Evolution, and Systematics* 46: 99–120.

Maggioni E (2015) Water demand management in times of drought: What matters for water conservation. *Water Resources Research* 51: 125–139.

Maslin M (2016) Forty years of linking orbits to ice ages. *Nature* 540: 208–209.

Maslin M (2020) Tying celestial mechanics to Earth's ice ages. *Physics Today* 73: 48–53.

Mason THE, Stephens PA, Gilbert G, Green RE, Wilson JD, Jennings K, Allen JRM, Huntley B, Howard C, Willis SG (2021) Using indices of species' potential range to inform conservation status. *Ecological Indicators* 123: 107343.

McCrory PA, Wilson DS, Stanley RG (2009) Continuing evolution of the Pacific–Juan de Fuca–North America slab window system—A trench–ridge–transform example from the Pacific Rim. *Tectonophysics* 464: 30–42

Meldahl KH (2015) *Surf, Sand, and Stone*. University of California Press, Oakland, CA, pp. 12–14.

Messner S, Miranda SC, Young E, Hedge N (2011) Climate-change related impacts in the San Diego region by 2050. *Climatic Change* 109 (Suppl 1): S505–S531.

Mitchell RN, Kilian TM, Evans DAD (2012) Supercontinent cycles and the calculation of absolute palaeolongitude in deep time. *Nature* 482: 208–212.

O'Gorman PA (2015) Precipitation extremes under climate change. *Current Climate Change Reports* 1:49–59.

Olsen MJ, Johnstone E, Ashford SA, Driscoll N, Hsieh TJ, Kuester F (2008) Rapid response to seacliff erosion in San Diego, California using terrestrial LIDAR.

Proceedings of the ASCE Solutions to Coastal Disasters 2008, Oahu, Hawaii, 573–583.

Olsen MJ, Johnstone E, Driscoll N, Kuester F, Ashford SA (2016) Fate and transport of seacliff failure sediment in Southern California. *Journal of Coastal Research* 76: 185–199.

Putirka K, Busby C (2011) Introduction—origin and evolution of the Sierra Nevada and Walker Lane. *Geosphere* 7: 1269–1272.

Rahmstorf S (2007) A semi-empirical approach to projecting future sea-level rise. *Science* 315: 368–370.

Riordan EC, Rundel PW (2014) Land use compounds habitat losses under projected climate change in a threatened California ecosystem. *PLoS One* 9(1): e86487.

Rosenberg KV, Dokter AM, Blancher PJ, Sauer JR, Smith AC, Smith PA, Stanton JC, Panjabi A, Helft L, Parr M, Marra PP (2019) Decline of the North American avifauna. *Science* 366: 120–124.

Sahakian V, Bormann J, Driscoll N, Harding A, Kent G, Wesnousky S (2017) Seismic constraints on the architecture of the Newport-Inglewood/Rose Canyon fault: Implications for the length and magnitude of future earthquake ruptures. *JGR Solid Earth* 122: 2085–2105.

SANDAG (2010) *2050 Regional Growth Forecast Process and Model Documentation.* San Diego Association of Governments, San Diego, CA.

Shields CA, Kiehl JT (2016) Atmospheric river landfall-latitude changes in future climate simulations. *Geophysical Research Letters* 43: 8775–8782.

Swain DL, Langenbrunner B, Neelin D, Hall A (2018) Increasing precipitation volatility in twenty-first century California. *Nature Climate Change* 8: 427–433.

Syphard AD, Bar Massada A, Butsic V, Keeley JE (2013) Land use planning and wildfire: Development policies influence future probability of housing loss. *PLoS ONE* 8(8): e71708.

Takekawa JY, Throne KM, Buffington KJ, Spragens KA, Swanson KM, Drexler JZ, Schoellhamer DH, Overton CT, Casazza ML (2013) Final report for sea-level rise response modeling for San Francisco Bay estuary tidal marshes. Open-File Report 2013-1081. *US Geologic Survey*. https://pubs.er.usgs.gov/publication/ofr20131081.

Tilman D, May R, Lehman CL, Nowak MA (1994) Habitat destruction and the extinction debt. *Nature* 371: 65–66.

UCS (2018) Underwater. Rising Seas, Chronic Floods, and the Implications for US Coastal Real Estate. Union of Concerned Scientists. Retrieved from https://www.ucsusa.org/sites/default/files/attach/2018/06/underwater-analysis-full-report.pdf; accessed February 1, 2021.

Ullrich PA, Xu Z, Rhoades AM, Dettinger MD, Mount JF, Jones AD, Vahmani P (2018) California's drought of the future: A midcentury recreation of the exceptional conditions of 2012–2017. *Earth's Future* 6: 1568–1587.

Voosen P (2020) A muddy legacy. *Science* 369: 898–901.

Wake DB, Vredenburg VT (2008) Are we in the midst of the sixth mass extinction? A view from the world of amphibians. *Proceedings of the National Academy of Sciences* 105: 11466–11473.

Wells ML, O'Leary JF, Franklin J, Michaelsen J, McKinsey DE (2004) Variations in a regional fire regime related to vegetation type in San Diego County, California (USA). *Landscape Ecology* 19: 139–152.

Wesnousky SG (2005) Active faulting in the Walker Lane. *Tectonics* 24: TC3009

Westerling AL, Cayan DR, Brown TJ, Hall BL, Riddle LG (2004) Climate, Santa Ana winds and autumn wildfires in Southern California. *EOS* 85: 289–300.

Young AP, Olsen MJ, Ddriscoll N, Flick RE, Gutierrez R, Guza RT, Johnstone E, Kuester F (2010) Comparison of airborne and terrestrial Lidar estimates of seacliff erosion in Southern California. *Photogrammetric Engineering & Remote Sensing* 76: 421–427.

Zauscher MD, Wang Y, Moore MJK, Gaston CJ, Prather KA (2013) Air quality impact and physicochemical aging of biomass burning aerosols during the 2007 San Diego wildfires. *Environmental Science & Technology* 47: 7633–7643.

Zhang KQ, Douglas BC, Leatherman SP (2004) Global warming and coastal erosion. *Climatic Change* 64: 41–58.

Index

www.ingramcontent.com/pod-product-compliance
Ingram Content Group UK Ltd.
Pitfield, Milton Keynes, MK11 3LW, UK
UKHW022158190926
475702UK00003B/94
* 9 7 8 1 1 3 8 5 9 6 7 6 4 *